Springer Theses

Recognizing Outstanding Ph.D. Research

Aims and Scope

The series "Springer Theses" brings together a selection of the very best Ph.D. theses from around the world and across the physical sciences. Nominated and endorsed by two recognized specialists, each published volume has been selected for its scientific excellence and the high impact of its contents for the pertinent field of research. For greater accessibility to non-specialists, the published versions include an extended introduction, as well as a foreword by the student's supervisor explaining the special relevance of the work for the field. As a whole, the series will provide a valuable resource both for newcomers to the research fields described, and for other scientists seeking detailed background information on special questions. Finally, it provides an accredited documentation of the valuable contributions made by today's younger generation of scientists.

Theses are accepted into the series by invited nomination only and must fulfill all of the following criteria

- They must be written in good English.
- The topic should fall within the confines of Chemistry, Physics, Earth Sciences, Engineering and related interdisciplinary fields such as Materials, Nanoscience, Chemical Engineering, Complex Systems and Biophysics.
- The work reported in the thesis must represent a significant scientific advance.
- If the thesis includes previously published material, permission to reproduce this must be gained from the respective copyright holder.
- They must have been examined and passed during the 12 months prior to nomination.
- Each thesis should include a foreword by the supervisor outlining the significance of its content.
- The theses should have a clearly defined structure including an introduction accessible to scientists not expert in that particular field.

More information about this series at http://www.springer.com/series/8790

Ada Altieri

Jamming and Glass Transitions

In Mean-Field Theories and Beyond

Doctoral Thesis accepted by
the Sapienza University of Rome, Rome, Italy
and Université Paris-Sud, Orsay, France

 Springer

Author
Dr. Ada Altieri
Département de Physique
École Normale Supérieure
Paris, France

Supervisors
Prof. Giorgio Parisi
Sapienza University of Rome
Rome, Italy

Prof. Silvio Franz
Centre scientifique d'Orsay
Université Paris-Sud
Orsay, France

ISSN 2190-5053   ISSN 2190-5061   (electronic)
Springer Theses
ISBN 978-3-030-23602-1   ISBN 978-3-030-23600-7   (eBook)
https://doi.org/10.1007/978-3-030-23600-7

This Springer imprint is published by the registered company Springer Nature Switzerland AG
The registered company address is: Gewerbestrasse 11, 6330 Cham, Switzerland

Modern physics: a path with a heart?
A path is only a path, and there is no affront,
to oneself or to others, in dropping it if that is
what your heart tells you. Look at every path
closely and deliberately. Try it as many times
as you think necessary. Then ask yourself
alone, one question. Does this path have a
heart? If it does, the path is good; if it doesn't
it is of no use.
Carlos Castaneda, *The Teachings of Don Juan*

Ai miei genitori

Supervisors' Foreword

The Ph.D. work of Dr. Ada Altieri was carried out at the Physics Department of the University of Rome "Sapienza" and at the Laboratoire de Physique Theorique et Modéles Statistiques (LPTMS) in Orsay.

The thesis concerns the definition and the study of different models of disordered systems which allow for an analytical description of the liquid-glass transition and the low-temperature vitreous phase.

The thesis opens with a detailed and pedagogical introduction to glassy systems in its phenomenological and theoretical aspects and a discussion of the open problems in the field. The original research work concerns two main themes: (1) the study of jamming and soft glassy phases within mean-field theory; (2) the study of systematic expansions allowing to go beyond mean field.

In the first part, the author focuses on several problems, in particular, (i) the jamming transition in the spherical perceptron. This model, that originates in the domain of artificial neural networks, in a modified version provides a high-dimensional version of a Lorentz gas and allows for a simple description of the universal features of the jamming transition; (ii) the generalization of the perceptron analysis to models of high-dimensional soft and hard spheres through the definition of an effective thermodynamic potential; and (iii) the analytical study of low-energy excitations in the transport properties of glasses and the determination of new critical exponents for the spectral density.

As a complement to the study of glasses, the thesis discusses models of dynamical evolution in ecosystems. Using a mechanical analogy, the author develops the idea that ecosystems can be driven to *marginally stable* and *jammed* states by their internal dynamics. In the last chapter, the author then discusses jamming soft modes in MacArthur's model of an ecosystem and highlighted interesting connections with complex landscapes in glasses.

The second part of the thesis is devoted to a beyond-mean-field investigation with detailed computations of finite-size and finite-dimension corrections in lattice theories. An *M-layer construction* is developed, allowing the definition of models that interpolate smoothly between finite-dimensional lattices and the exactly solvable Cayley tree, where the *Bethe approximation* is exact. The author shows how a systematic diagrammatic expansion can be established around the Bethe solution and how the resulting series can be re-summed near a critical point. Indeed, the critical series defined for a generically observable can be re-expressed as a sum of appropriate diagrams with the same prefactors of ordinary field theories.

This new approach promises important advances in the study of percolation problems, zero-temperature systems, but also of phase transitions in disordered systems such as the Random Field Ising Models that in finite dimension exhibit a different phenomenology from the mean-field case.

The value of this thesis is not limited to new and original results, all published in peer-reviewed papers, but it is also enriched by the versatility of the author to explore different fields and provide transverse perspectives.

Rome, Italy
Orsay, France
April 2019

Prof. Giorgio Parisi
Prof. Silvio Franz

Abstract

The detailed description of disordered and glassy systems represents an open problem in statistical physics and condensed matter. As yet, there is no single, well-established theory allowing to understand such systems. The research presented in this thesis is related in particular to the study of glassy materials in the low-temperature regime. More precisely, in systems formed by athermal particles subject to repulsive short-range interactions, upon progressively increasing the density, a so-called jamming transition can be detected. It entails a freezing of the degrees of freedom and hence a huge increase of the material rigidity.

We shall study this problem in view of a formal analogy between sphere models and the perceptron, a theoretical model undergoing a jamming transition and frustration phenomena typical of disordered systems. Being a mean-field model, it allows to obtain exact analytical results, which are generalizable to more complex high-dimensional settings.

The main aim is to reconstruct the vibrational spectrum and all relevant properties of a specific phase of the perceptron, corresponding to the hard-sphere regime. In this framework, we will derive the effective potential as a function of the gaps and forces between particles, and we will show that it is dominated by a non-trivial logarithmic interaction near the jamming point. This interaction in turn will clarify the relations existing between the relevant variables of the system, in the critical jamming region and beyond.

Understanding the jamming transition and the perceptron properties will allow us to make progress in several related fields. First, this study could lay part of the groundwork toward a complete theory of amorphous systems, in both infinite and finite dimensions. Furthermore, the perceptron model seems to have a close connection with the so-called von Neumann problems. Indeed, biological and ecological systems often develop pseudo-critical properties and give rise to general mechanisms of resource-consumption optimization. Is the identification of a broken symmetry regime possible? What would it yield in terms of the spectrum of energy fluctuations? These are just a few questions we shall attempt to answer in this context.

However, the mean-field approximation can sometimes provide wrong or mis-leading information, especially in studying certain phase transitions and deter-mining the exact lower and upper critical dimensions. To have a broader perspective and correctly deal with finite-dimensional systems, in the second part of the thesis, we will discuss how obtaining a systematic perturbative expansion which can be applied to any model, as long as defined on a lattice or a bipartite graph. Our motivation is in particular due to the possibility of studying relevant second-order phase transitions which exist on the Bethe lattice—a lattice with a locally tree-like structure and fixed connectivity for each node—but which are either qualitatively different or absent in the corresponding fully connected version.

Publications Related to This Thesis

- A. Altieri, S. Franz
  Constraint satisfaction mechanisms for marginal stability and criticality in large ecosystems,
  Rapid Communications in Phys. Rev. E **99**, 010401(R) (2019).

- A. Altieri,
  Higher order corrections to the effective potential close to the jamming transition in the perceptron model,
  Phys. Rev. E **97**, 012103 (2018).

- A. Altieri, M. C. Angelini, C. Lucibello, G. Parisi, F. Ricci-Tersenghi, T. Rizzo,
  Loop expansion around the Bethe approximation through the M-layer construction,
  J. Stat. Mech. (2017) 113303.

- A. Altieri, S. Franz, G. Parisi,
  The jamming transition in high dimension: an analytical study of the TAP equations and the effective thermodynamic potential,
  J. Stat. Mech. (2016) 093301.

- A. Altieri, G. Parisi, T. Rizzo,
  Composite operators in cubic field theories and link overlap fluctuations in spin-glass models,
  Phys. Rev. B **93**, 024422 (2016).

Acknowledgements

This doctoral thesis has been developed in a co-tutorship project between the Universities Sapienza in Rome and Paris-Sud XI in Orsay, under the supervision of Giorgio Parisi and Silvio Franz, respectively. First, I would like to warmly thank my two supervisors to whom I am indebted for their continuous support and for numerous insightful discussions during these 3 years. To Giorgio, already supervising my master thesis, my greatest thanks for having given me the opportunity to work with him and introduced me to scientific research. Every single computation has given rise to new and always stimulating discussions. My warmest thanks likewise to Silvio, for his great guiding and incomparable teaching role, day by day. I will never cease to be thankful to Giorgio and Silvio, for the opportunity to learn and grow up in a rich and versatile environment. They always transmitted me their enthusiasm and passion for physics, trying to investigate and to discover together new physical phenomena. It was a pleasure but especially a great honor for me to work with these two scientists. I will always carry with me the memories of this experience: our discussions, my questions, my doubts, but also the joy of seeing a project started together finally realized. I really feel the French experience changed my mind, making me safer and more geared to face everyday challenges.

In this perspective, I wish to warmly thank Francesco Zamponi and Pierfrancesco Urbani for very stimulating discussions and their precious advice during the Ph.D. and the first part of my post-doctoral activity in collaboration with them. Moreover, I thank all scientists that I had the chance to meet and discuss with, in Italy, France, U.S., and in all parts of the world I set foot on, especially Enzo Marinari who let me appreciate statistical mechanics and chaotic systems already in the first months of my bachelor program and who was a constant source of attentive suggestions over the years; Tommaso Rizzo who co-supervised my master thesis with Giorgio and who was a reference point for clarifying any curiosity on the most difficult mathematical structures; Giulio Biroli whose fascinating papers inspired and introduced me to the study of glasses during the first year of my Ph.D. and with whom it is always very illuminating to discuss, starting with an amazing Beg Rohu Summer School and going on during my current postdoc; Matthieu Wyart for his important point of view on some results described in this thesis, and moreover the director of the LPTMS Emmanuel Trizac, Maria Chiara Angelini, Andrea Cavagna, Eric DeGiuli, Luca Leuzzi, Carlo Lucibello, Federico Ricci-Tersenghi, and Pierpaolo Vivo.

Then, I strongly thank my two external referees, Patrick Charbonneau and Grzegorz Szamel, for their interest and their time devoted to my research project, as well as the members of the committee, Giulio Biroli, Irene Giardina, Enzo Marinari, and Guilhem Semerjian, for having willingly accepted my invitation to be present on the final, important day of my doctoral studies.

Many thanks to my friends and colleagues in the Simons office and abroad, in particular, Elisabeth Agoritsas, Marco Baity-Jesi, Silvia Grigolon, Beatriz Seoane, and Gabriele Sicuro. I cannot forget special thanks to my greatest friends, Simone and Paolo, who stood me since fall 2009, and Silvia, one of my precious source of advice, a wonderful person to speak with, laugh, cry, and to count on, always.

I am very thankful to the Laboratoire de Physique Théorique et Modèles Statistiques where part of this thesis has been realized, for a welcoming and nice reception since my second year of Ph.D. Then, special attention is devoted to Sapienza University where this thesis was defended and which has been an anchorage since the beginning of this journey, 8 years ago.

Moreover, I acknowledge the Université Franco-Italienne/Università Italo-Francese (UFI-UIF) that additionally funded my research and let me increase my network of collaborations, in Italy and France. I want also to express my gratitude to the Simons Collaboration on "Cracking the glass problem", funded by the Simons Foundation, giving me the chance to interact with a vast community of physicists and to keep always up with scientific debates on glasses.

Finally, I want to say a few lines to thank my parents, for their love, their support, and for the huge sacrifices they did for me, in terms of distance and everything. I am perfectly aware that I would never have achieved these important goals without them. Despite our different interests and passions, they always supported me in my choices and gave me the motivation to insist and persist.

Mum, thanks for listening to my long discussions about physics without understanding anything, but still trying to show interest and participation. Mum, dad, thank you to let me follow my dreams and live the life I have always wanted to, although far away from your place.

Contents

Acronyms

BP	Bethe–Peierls or Belief Propagation
CRR	Cooperative Rearranging Region
CSP	Constraint Satisfaction Problem
DOS	Density Of States
EMT	Effective Medium Theory
ERG	Erdös-Rényi Graph
FA	Fredrickson-Andersen
FC	Fully Connected
HNC	Hyper-netted chain
IPR	Inverse Participation Ratio
KA	Kob-Andersen
KCM	Kinetically Constrained Model
MCT	Mode-Coupling Theory
NBW	Non-Backtracking Walk
RCP	Random Close Packing
RFIM	Random Field Ising Model
RFOT	Random First-Order Transition
RRG	Random Regular Graph
RS	Replica Symmetry
1RSB	One-step Replica Symmetry Breaking
RSB	Replica Symmetry Breaking
SG	Spin Glass
TAP	Thouless-Anderson-Palmer
VFT	Vogel-Fulcher-Tamman
ZRP	Zero-Range Process

Chapter 1
Introduction

This thesis is devoted to the study of disordered systems, both in infinite and in finite dimensions. What the author had in mind while writing this manuscript was to deal with something spectacular, which in every single concept tickled her attention and curiosity. A curiosity arisen from the study of spin glasses and slowly flowed into structural glasses at large, with all the implications that followed. This thesis is actually developed in this spirit, aiming to underline interesting, exotic, sometimes bizarre connections between these two sides of the same coin. The goal is to unveil the dense network of connections existing between glasses, granular materials, ecosystems, metabolic networks and beyond. We will attempt to touch upon several of these topics, ranging from optmization theory to optimal packings of spheres, to real-space condensation, both in low and high dimensions, as well as to random matrices and lattice theories. The common thread in all these fields is the disorder.

The most paradigmatic example of disordered system dates back to Edwards and Anderson [8], who were followed by Sherrington and Kirkpatrick [28] in proposing a simplified model of spin glass (SG) in its fully-connected version (FC). A spin glass is essentially a magnetic system characterized by purely random interactions between the spin variables, which in turn generate frustration and complex dynamics [9]. Even from a mean-field perspective, the above-mentioned model exhibits a rich and intriguing phenomenology. Several efforts came in succession. Among them it is worth mentioning Parisi's solution based on the *replica method* [25, 26], which is a cornerstone for a vast class of complex systems.

The simplest SG model consists of N spins σ_i interacting via pairwise interactions defined by the couplings J_{ij}. For simplicity we assume the spins to be Ising variables, i.e. $\sigma_i = \pm 1$, and the couplings to be symmetric variables, i.e. $J_{ij} = J_{ji}$ and with zero diagonal contributions. The Hamiltonian of the system reads:

$$\mathcal{H}_{\text{SG}} = -\sum_{i,j}^{1,N} J_{ij}\sigma_i\sigma_j \,, \tag{1.1}$$

© Springer Nature Switzerland AG 2019
A. Altieri, *Jamming and Glass Transitions*, Springer Theses,
https://doi.org/10.1007/978-3-030-23600-7_1

where the J's are extracted from a given probability distribution, usually a Gaussian distribution. The couplings are *quenched* variables meaning that their fluctuations, compared to the fluctuations of the spins, are negligible.

The model is defined on a lattice, where the spins correspond to the nodes and the random couplings to the links between two nodes. Although this model appears to be rather simple, it describes features that belong to a wide class of different systems. Examples are given by the Sherrington-Kirkpatrick model (namely the fully-connected version), the Bethe lattice, the Edwards-Anderson one (essentially a finite-dimensional variant) and the large-range models. All these systems were found to be equivalent in the high-connectivity or infinite-dimension limit. In this regime, the aforementioned systems reduce to the Sherrington-Kirkpatrick model.

Disordered systems are ubiquitous and they encompass very different fields. Irrespective of this versatile scenario two main universality classes can be identified: the first, including the models mentioned above (other than eventually the finite-dimensional one), is characterized by a continuous transition from a paramagnetic phase to a spin-glass disordered phase (a.k.a. a *full replica symmetry breaking* description) [24, 26]; the second instead displays a discontinuous transition (a *one-step replica symmetry breaking* (1RSB) description)- also refereed to as Random First Order Transition (RFOT) [5, 13, 14]- and includes fragile glasses and several optmization problems. The simplest and most representative model that belongs to this second class is the p-spin with $p \geq 3$:

$$\mathcal{H} = - \sum_{i_1,\ldots,i_p}^{1,N} J_{i_1,\ldots,i_p} \sigma_{i_1} \sigma_{i_2} \ldots \sigma_{i_p} , \tag{1.2}$$

where the couplings J's are symmetric under permutations of the indices and zero if at least two indices are equal.

However, this classification was upset in the eighties when Gardner made a new proposal [12]. Deeply into the glass phase, RFOT systems can undergo another phase transition, the so-called *Gardner transition*. In correspondence of it, each amorphous state, say a basin, is fragmented into a full fractal structure of sub-basins. Therefore, the internal structure of states in which a basin splits is described by the full RSB [26] solution of the partition function. Analytical computations in hard-sphere systems in the infinite dimension limit also predict this type of transition favoring the emergence of a fractal hierarchy [7]. The infinite-dimensional limit theory thus corroborates the hypothesis that an equilibrium glass state modifies its structure from a normal to a marginal glass phase, at a finite pressure. The concept of *marginality* is related to the appearance of soft vibrational modes at low-frequency scales [30, 31]. By compressing the system in this regime, the endpoint of the compression line corresponds to the *jamming* onset, where the pressure diverges. Thus the jamming transition takes place inside the glass phase described by a complex energy landscape within a full RSB picture as well.

This thesis is divided into two parts. The first part concerns the physics of glasses in the low-temperature regime with a special emphasis on the jamming transition.

To better investigate this phenomenon we make use of the knowledge of continuous constraint satisfaction problems (CSPs) and their fundamental and, at the same time, amazing connection with sphere models in high dimensions. Recently, new interesting results have been achieved once it was realized that the perceptron model, borrowed from neural networks but re-proposed here in a modified form, falls in the same universality class as hard spheres in infinite dimensions [10, 11].

Our main goal is the derivation of an effective thermodynamic potential (or Thouless-Anderson-Palmer free energy (TAP) [29]) in this exactly solvable model. Our approach bridges the gap between different scenarios, in particular between the dynamical mode-coupling-like equations [15, 17] and the replica method [7, 16]. We perform a formal perturbative expansion of the free energy which turns out to be a function of both the generalized forces and the gaps between particles. The clear advantage of this method is the possibility to study in detail the hard-sphere regime of the perceptron model and to reconstruct the spectrum of small harmonic vibrations, both in the *liquid*—replica symmetric phase—and in the *marginal glass*—replica symmetry breaking—phase. We highlight a different behavior with respect to the soft-sphere regime with a non-trivial exponent exactly derivable in this framework. Moreover, we prove the emergence of a logarithmic interaction in the small-gap regime of the perceptron phase diagram, as was predicted first numerically in [6] for three-dimensional glasses. This constitutes a pivotal outcome which corroborates the hypothesis of an underlying universality, irrespective of the microscopic details and the specific dimension of the system. The definition of a suitably coarse-grained potential also provides the tools to analyze higher-order corrections to the TAP free energy and to make predictions about relevant scaling regimes close to the jamming line.

Jamming is then the focal point of the first part of this thesis. In fact, this phenomenon has a broader spectrum of applicability than one might expect at first sight. It does not concern the domain of structural glasses only, but also error correcting codes, traffic flow, transport in crowded biological environments and evolutionary dynamics. Another goal of this thesis is indeed to establish a direct connection between generic optmization problems and models of great interest in ecology or biophysics. The common thread with the jamming onset is the existence of a mechanism of marginal mechanical stability [18, 21–23, 27], which can be also recovered in generic Von Neumann problems.

All formalisms developed in the first part deal with fully-connected models: the continuous perceptron is the key example. It can nevertheless appear rather reductive. Then, to cover a broad spectrum of physical phenomena and extend our predictions beyond mean field, in the second part we mostly focus on the computation of finite-size corrections in lattice theories. Fully-connected mean-field models are obviously a good and reasonable starting point to perform a perturbative expansion. However, as soon as one attempts to extend these results to finite dimensions, they turn out to be not really appropriate to capture all relevant features. Alternative approaches can be considered, for instance by using diluted random graphs. In the thermodynamic limit, diluted models exhibit a tree-like structure locally, which allows to safely use the cavity method [20, 26] (also referred to as *Bethe approximation* in the following). However, in a diluted system of finite size, strong deviations from the Bethe approximation might occur accompanied by uncontrolled renormalization schemes and diverging perturbative theories. In the last part of this thesis we address this problem, connecting finite-size corrections to the presence of short loops in a graph. The method is based on the definition of a specific multi-layer construction on lattice models, in which a perturbative expansion around the Bethe solution can be performed.

Structure of the Thesis

- In more details, in Chap. 2 we introduce the reader to the phenomenology behind supercooled liquids and the glass phase. The first part is devoted to the explanation of some of the most commonly used dynamical descriptions—such as Mode-Coupling Theory and Dynamical Facilitation—while the second one is mainly focused on static approaches.
- In Chap. 3 we describe the jamming transition as another peculiar feature of the glass phenomenology in the zero-temperature regime. We start from a simple general description pictured in the phase diagram initially introduced by Nagel and Liu [19]. Then we present the main open questions related to anomalous modes, critical distributions at jamming, diverging correlation lengths.
- In Chap. 4 we address some aspects concerning the jamming transition in high dimensions. We introduce the recently highlighted connection between a continuous constraint satisfaction problem, the perceptron in the non-convex phase, and sphere models. This similarity allows us to derive a unifying mean-field formalism and then to investigate the jamming and unjamming regimes.
- In Chap. 5 we derive an analogous effective potential in high-dimensional sphere models from which we can make predictions about critical phases. This approach is a starting point for addressing possible *condensation* phenomena in real space.
- In a broader perspective, Chap. 6 is devoted to obtaining a deeper understanding of the relationship between the perceptron model in its convex phase and a resource-competition model in ecosystems. We show the analogies between these frameworks, looking at the phase diagram and the vibrational spectrum and focusing on the emergence of what would correspond to a marginally stable phase, in the parlance of statistical physics.
- To extend these results beyond mean field, in Chap. 7 we propose a general method, based on a multi-layer construction. This approach allows us to obtain a systematic perturbative expansion around the Bethe free energy, valid both in the ordered and the disordered phases. Varying the tunable parameter M, which defines the number of layers piled up on the original lattice model, we are able to interpolate between the finite-dimensional original graph ($M = 1$) and the locally tree-like graph wherein the Bethe approximation is correct ($M \to \infty$).

Most of the results in this thesis have been published, although the thesis offers a more detailed presentation. Chapters 2 and 3 attempt to make the present thesis self-contained and to give a broad overview of glasses. Original results are contained in Chaps. 4– 7:

- Chapter 4 refers to [1, 3];
- Chapter 5 refers to [3] just for the first part, while the second part contains unpublished results;
- Chapter 6 refers to [2];
- Chapter 7 refers to [4].

References

1. Altieri A (2018) Higher-order corrections to the effective potential close to the jamming transition in the perceptron model. Phys Rev E 97(1):012103
2. Altieri A, Franz S (2019) Constraint satisfaction mechanisms for marginal stability and criticality in large ecosystems. Phys Rev E 99(1):010401(R)
3. Altieri A, Franz S, Parisi G (2016) The jamming transition in high dimension: an analytical study of the TAP equations and the effective thermodynamic potential. J Stat Mech: Theory Exp 2016(9):093301
4. Altieri A et al (2017) Loop expansion around the Bethe approximation through the M-layer construction. J Stat Mech: Theory Exp 2017(11):113303
5. Biroli G, Bouchaud JP (2012) The random first-order transition theory of glasses: a critical assessment. In: Wolynes PG, Lubchenko V (eds) Structural glasses supercooled liquids: theory, experiment and applications. Wiley
6. Brito C, Wyart M (2006) On the rigidity of a hard-sphere glass near random close packing. Europhys Lett 76(1):149
7. Charbonneau P et al (2014) Fractal free energy landscapes in structural glasses. Nat Commun 5:3725
8. Edwards SF, Anderson PW (1975) Theory of spin glasses. J Phys F Metal Phys 5(5):965
9. Fischer KH, Hertz JA (1991) Spin glasses (Cambridge Studies in Magnetism)
10. Franz S, Parisi G (2016) The simplest model of jamming. J Phys A Math Theor 49(14):145001
11. Franz S et al (2015) Universal spectrum of normal modes in low-temperature glasses. Proc Natl Acad Sci 112(47):14539
12. Gardner E (1985) Spin glasses with p-spin interactions. Nucl Phys B 257:747
13. Kirkpatrick TR, Thirumalai D, Wolynes PG (1989) Scaling concepts for the dynamics of viscous liquids near an ideal glassy state. Phys Rev A 40(2):1045
14. Kirkpatrick TR, Wolynes PG (1987) Stable and metastable states in mean-field Potts and structural glasses. Phys Rev B 36(16):8552
15. Kurchan J, Maimbourg T, Zamponi F (2016) Statics and dynamics of infinite dimensional liquids and glasses: a parallel and compact derivation. J Stat Mech: Theory Exp 2016(3):033210
16. Kurchan J, Parisi G, Zamponi F (2012) Exact theory of dense amorphous hard spheres in high dimension. I. The free energy. J Stat Mech: Theory Exp 2012(10):P10012
17. Kurchan J et al (2013) Exact theory of dense amorphous hard spheres in high dimension. II. The high density regime and the gardner transition. J Phys Chem B 117:12979
18. Le Doussal P, Müller M, Wiese KJ (2010) Avalanches in mean-field models and the Barkhausen noise in spin-glasses. EPL (Europhys Lett) 91(5):57004
19. Liu AJ, Nagel SR (1998) Nonlinear dynamics: jamming is not just cool any more. Nature 396(6706):21
20. Montanari A, Mézard M (2009) Information, physics and computation. Oxford University Press, Oxford
21. Müller M, Pankov S (2007) Mean-field theory for the three-dimensional Coulomb glass. Phys Rev B 75:144201
22. Müller M, Wyart M (2015) Marginal stability in structural, spin, and electron glasses. Ann Rev Condens Matter Phys 6:177

23. Pankov S, Dobrosavljević V (2005) Nonlinear screening theory of the Coulomb glass. Phys Rev Lett 94(4):046402
24. Parisi G (1980) A sequence of approximated solutions to the SK model for spin glasses. J Phys A Math General 13(4):L115
25. Parisi G (1997) On the replica approach to glasses. arXiv preprint arXiv:condmat/9701068
26. Parisi G, Mézard M, Virasoro MA (1987) Spin glass theory and beyond. World Scientific, Singapore
27. Pollak M (1970) Effect of carrier-carrier interactions on some transport properties in disordered semiconductors. Discuss Faraday Soc 50:13
28. Sherrington D, Kirkpatrick S Solvable model of a spin-glass. Phys Rev Lett 35(26):1792
29. Thouless DJ, Anderson PW, Palmer RG (1977) Solution of 'solvable model of a spin glass'. Philos Mag 35(3):593
30. Wyart M (2012) Marginal stability constrains force and pair distributions at random close packing. Phys Rev Lett 109:125502
31. Wyart M (2005) On the rigidity of amorphous solids. Ann Phys 30:1

Part I
Glass and Jamming Transitions
in Mean-Field Models

Chapter 2
Supercooled Liquids and the Glass Transition

The origin of the glass transition is one of the most debated, open problems both in experimental and theoretical physics. Already in the nineties the Nobel Prize winner P. W. Anderson asserted that *the deepest and most interesting unsolved problem in solid state theory is probably the theory of the nature of glass and the glass transition.*

Numerous studies have been carried out over the last twenty years but an established and unequivocal theoretical framework has not yet completely been achieved. The qualifier *glassy* is used nowadays in a wide variety of contexts to describe systems with a slow, sluggish dynamics whose degrees of freedom are not apparently ordered. Their amorphous behavior can be either self-generated or due to the presence of quenched impurities and defects in the system.

Essentially, a glass can be described as a liquid subject to a huge slowing down of the diffusive motion of the degrees of freedom. It is accompanied by a spectacular increase of the relaxation time and of the viscosity, up to 14 orders of magnitude for a 30% variation in temperature. However, this transition does not correspond to a true transition according to Ehrenfest's definition. The glass transition temperature T_g is only empirically defined as the threshold below which the material becomes too viscous to have a diffusive motion on an experimentally relevant time scale. Another aspect of the glass transition to be addressed carefully is the presence of the crystal: the quenching process must be fast enough in order to avoid the transition to a crystalline phase. Once the existence of the liquid is guaranteed against crystallization, the process of cooling down the system with a typical cooling rate of 0.1–100 K/min, drives the system into a *supercooled liquid phase*. At the glass transition temperature, T_g, the system is unavoidably out of equilibrium with non-trivial amorphous properties. Although many features of the glassy phase can also be found in several liquids, the physics behind the glass transition is really different and cannot be simply described in the context of equilibrium phenomena. The main qualitative features characterizing a glass are: (i) a non-exponential time dependence of the relaxation functions, leading also to a non-Debye behavior of the frequency-dependent susceptibilities; (ii) the emergence of several relaxation regimes; (iii) the

© Springer Nature Switzerland AG 2019
A. Altieri, *Jamming and Glass Transitions*, Springer Theses,
https://doi.org/10.1007/978-3-030-23600-7_2

presence of dynamical heterogeneities and (iv) a dramatic decrease of the configurational entropy, which properly marks the difference between the entropy of a crystal and that of a supercooled liquid.

Several protocols can be pursued to obtain a glass. The choice usually depends on the material and range from a fast cooling of the liquid to decompression of crystals at high pressure, chemical reactions, evaporation of solvents, etc. To give an idea of the complexity of the glass transition, one can imagine following a liquid, prepared at enough high temperature, during a cooling. In the warm liquid several processes occur on different timescales. Upon cooling the liquid below the melting point T_m, the shear relaxation time and the viscosity grow very sharply. This appears as a completely general property of many supercooled liquids, including polymers.

Cooling the material down even more, at the *glass temperature* T_g the system enters a phase where its relaxation time grows so sharply to exceed any experimental accessible time. In this phase, the system cannot even reach any new equilibrium state. This peculiar behavior gives a phenomenological definition of the glass transition, which is actually detached from any connection with a thermodynamic transition in the usual sense. Indeed, the ergodicity breaking occurring in a glass has a completely different nature compared to a crystal, a dynamical phenomenon in the first case and a truly thermodynamic transition in the second. Since the glass transition temperature is defined in terms of the relaxation time, one might wonder whether it is a meaningful concept or just a speculative artifact. As shown in Fig. 2.1, different values of the experimental time t_{exp} correspond to different values of T_g. Nevertheless, the dependence on the experimental protocol affects the transition point only marginally: a change of order one in the experimental time results in a logarithmic change in the corresponding temperature. This argument is also corroborated by the Bartenev-Ritland phenomenological equation [4, 47], which relates the glass transition temperature to the cooling rate:

$$1/T_g = a - b \log(r) , \qquad r = \frac{dT}{dt} \qquad\qquad (2.1)$$

where a and b are two phenomenological constants inferred from experimental data.

Besides the glass transition temperature other remarkable physical processes deserve our attention, such as the crossover regime from non-activated to activated dynamics, which takes place at T_x and is known for spin-glass models as the *Goldstein temperature*.[1]

Going deeply into the supercooled phase and pretending to be able to avoid the glass transition and hence to preserve the liquid at equilibrium, the Kauzmann temperature, T_K, is finally reached. At this temperature the entropy of the liquid exactly coincides with that of the crystal. Let us try to understand why this aspect can be very delicate. The crucial point relies on the possibility to split the entropy of an

---

[1] Above T_x the dynamics is described by the Mode-Coupling theory (MCT), which we will present in more details in the following. The nature of this crossover can be better understood in the framework defined by the p-spin model and the MCT, both predicting a dynamical transition at $T_d \equiv T_c$, which is presumably the Goldstein temperature.

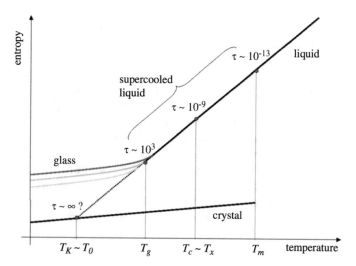

Fig. 2.1 Schematic plot of the entropy as a function of temperature in a liquid, ranging from the melting temperature T_m—where a first-order phase transition liquid/crystal occurs—to the deep supercooled liquid phase. Below T_m the liquid branch becomes metastable. T_c is the so-called Mode-Coupling temperature where a purely dynamic transition is located. This temperature will be also refereed in the text as T_d to denote its purely dynamical origin. T_x stands for the Goldstein temperature, determining a crossover between a high-temperature non-activated dynamics to a low-temperature one with activation processes. The last two relevant temperatures are T_g, corresponding to the glass transition, and T_K, the Kauzmann temperature, where the extrapolated liquid entropy coincides with the crystal entropy. Picture taken from [12]

equilibrium liquid into two terms, a vibrational contribution, taking into account the short-time vibrations in each minimum of the potential energy landscape, and a configurational one, due to the presence of multiple amorphous minima. The latter is different from zero only if the number of minima is exponentially large with respect to the size of the system, i.e. $S_c = \frac{1}{N} \log \mathcal{N}$, $\mathcal{N}$ being the number of minima at equilibrium. Let us assume the vibrational entropy around a minimum—a local minimum for the liquid and global one for the crystal respectively—to be roughly the same as in a crystal.[2] This approximation, in some cases considered too rough and ill-defined as, for instance, in hard spheres [44], leads to:

$$\Delta S(T) = S_{\text{liq}}(T) - S_{\text{cr}}(T) \sim S_c(T) , \qquad (2.2)$$

S_{liq} and S_{cr} being the liquid and the crystal entropy contributions respectively. Equation (2.2) highlights a fundamental property of supercooled liquids, namely the equality between the excess entropy and the configurational entropy $S_c(T)$. Since the configurational entropy of a supercooled liquid decreases sharply at low temperatures, the extrapolation would determine a point at which the entropy becomes smaller than

---

[2]Further contributions would be essentially sub-leading.

that of the crystal. To handle this issue,[3] Kauzmann claimed that for a huge class of systems the configurational entropy should vanish at a finite temperature T_K (in many other cases, e.g. in oxides, this temperature is exactly zero meaning that no transition occurs). A zero configurational entropy means that the system can visit only a finite, sub-exponential number of minima. Since the number of allowed configurations cannot further decrease and cannot become negative, the configurational entropy must be zero even in the temperature range $T < T_K$.[4]

Differently than the glass transition, at the Kauzmann temperature the system should undergo a true phase transition (also referred to as an *ideal glass transition*), signaled by a divergence of the relaxation time. Its definition would not need to invoke any experimental time scale. However, the existence of this transition has long been debated, wondering whether it can really occur in glass formers. The problem is that the dynamic glass transition, at T_g, prevents us to actually reach this point.

2.1 Arrhenius and Super-Arrhenius Behaviors

The relaxation time of a glass-forming liquid depends first and foremost on the temperature (or on the pressure) at which the experiments are performed. According to the Maxwell relation, it is proportional to the viscosity η:

$$\eta \sim G_g \tau_{eq} \, , \tag{2.3}$$

where G_g stands for the instantaneous shear modulus. Note that the elastic shear modulus, $G(t)$, plays an important role in defining the solidity of a material since it goes to zero at infinite times in a liquid and decays to a plateau in a solid. In the simplest approximation, $G(t)$ is provided by the Maxwell relation, namely:

$$G(t) = G_\infty \exp\left(-t/\tau_R\right) \, , \tag{2.4}$$

where τ_R stands for the shear relaxation time. If $G(s)$ for a generic time s decays to zero slower than $1/s$, the shear relaxation function is not integrable in the long-time regime. Applying a constant stress, it flows slower and slower with a shear modulus dependence $G(t) \sim 1/t^a$, with $a < 1$. Materials satisfying this relation are usually called *power-law fluids*.

As mentioned above, the glass transition temperature requires a hands-on definition. Therefore, according to [40], it can be useful to fix a conventional value of the maximum experimental time one can wait before the liquid reaches equilib-

---

[3]There is no law in nature that actually forbids this phenomenon. In the crystallization transition in hard-sphere systems, for instance, the crystalline entropy can be larger than that of the liquid at high density.

[4]Actually, Kauzmann never gave a name to what is known now as *the Kauzmann temperature* but he rather introduced the concept of a *kinetic spinodal* $T_{sp} > T_k$. Such a spinodal point defines a regime below which the supercooled liquid does not exist anymore.

rium, i.e. $\tau_R(T_g) \sim 10^2 - 10^3$ s. Putting this relation together with Eq. (2.3), one can immediately obtain the value of the viscosity corresponding to the transition:

$$\eta(T_g) \sim 10^{12} \text{ Pa s} \tag{2.5}$$

given that $G_\infty \sim 10^9 - 10^{10}$ Pa. The viscosity thus is a reference tool to identify the transition point. However, data normally spread over a vast range due to the different values of T_g in different systems. For this reason, it is more convenient to picture the logarithm of the viscosity as a function of T_g/T, in such a way that all curves have the same value (that is 10^{12} Pa s) at $T_g/T = 1$. Figure 2.2, conventionally called *Angell's plot*, can also provide useful information to identify the different type of glass formers. The straight line corresponds to *strong* glasses (including silica, GeO_2, B_2O_3), which are well-described by an Arrhenius law:

$$\tau = \tau_0 \exp\left(\frac{E}{k_B T}\right), \tag{2.6}$$

where E is the effective activation energy necessary to jump an energy barrier E. Others, which display a strong deviation from the straight line, are usually described by a sharpe super-Arrhenius law and are said to belong to the class of *fragile* materials. Hence, the fragility of a glass former can be measured by the slope of the logarithm of the viscosity as a function of T_g/T, in $T_g/T = 1$. A sharper growth is a signature of a much more fragile behavior.[5] However, beyond this *strong-fragile* classification, there is a broader spectrum of phenomenologies that occur between these two extremal cases, as we can clearly see in Fig. 2.2.

In this plethora of glass-former behaviors it is thus essential to understand pros and cons of each phenomenological explanation. More precisely, in the high-temperature regime most liquids are well described by an exponential law linking the relaxation time (or the viscosity, which is more easily accessible in experiments) with the temperature change, as shown in Eq. (2.6). To fit data in the intermediate as well as in the low-temperature phase, several interpretations have been proposed, one being the Vogel-Fulcher-Tamman law (VFT):

$$\tau = \tau_0 \exp\left(\frac{A}{T - T_0}\right). \tag{2.7}$$

It is essentially a three parameter (τ_0, A, T_0) fit formula, appropriate to interpolate among many different glass formers over a broad range of temperatures, for $T_g < T < T_c$. In particular the parameter T_0 allows to interpolate between a purely Arrhenius behavior and a more fragile one, the latter being described by an increasing value of $T_0 > 0$. An analysis of this relation suggests a correspondence between the Kauzmann temperature and the fitted parameter T_0. At $T = T_0$ the relaxation time

---

[5]The concept of fragility in a glass former must not be confused with a structural and visible property, but as a representation in the configurational space of the degrees of freedom.

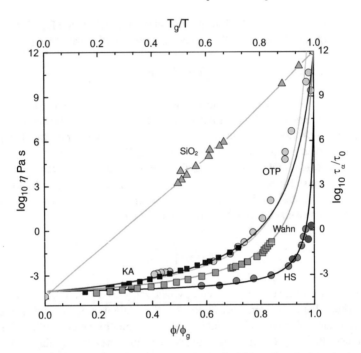

Fig. 2.2 Angell plot representing the logarithm of the viscosity versus the inverse temperature scaled by T_g for several glass-forming liquids [49]. For *strong* glasses the viscosity increases according to an Arrhenius law upon decreasing the temperature, while *fragile* liquids follow a super-Arrhenius behavior for which the activation energy grows as temperature decreases. For instance, SiO_2 has a viscosity of $2.4\,10^3$ Pa·s, about 300 degrees of freedom above its melting point. In the case of hard spheres the control parameter is the volume fraction, which is scaled to have $T_g/T = \phi/\phi_g$

diverges as a signature of the emergence of a critical transition. Despite the fact that no consensus exists on the deep physical meaning of the Kauzmann temperature and of the type of transition associated with it, for a wide class of systems these two temperatures coincide.

The VFT law is as well an object of debate. Criticism mostly concerns the inaccuracy of this law in describing sufficiently low-temperature regimes. Alternative descriptions are, for instance, based on the Bässler law:

$$\tau(T) = \tau_0 \exp\left(\frac{B}{T^2}\right) \tag{2.8}$$

where the exponent 2 can be obtained in different ways [5, 26, 31]. B again is just a fit parameter. A first difference between these scaling forms is the presence or the absence of a special temperature T_0, below which the system falls out of equilibrium, which determines a non-trivial glassy phase. For $T_0 = 0$, equilibrium can persist up to $T = 0$ without any glassy phase emerging.

A central question that might arise is whether a general theory of the glass transition is possible, irrespective of microscopic details and intrinsic differences from one material to another. Spin glasses provide a paradigmatic example of the opposite situation. Investigating in detail their phenomenology can be rather involved in some particular cases. For instance, there is no consensus about the existence of a transition in presence of an external magnetic field. However, despite some debated and open questions, a well-defined framework is known for spin-glass models. We cannot state the same for glass-forming liquids, for which both the choice of a specific minimal model as a good starting point, and the actual consistency of an effective theory of glasses are still matters of debate.

In the following we will attempt to present some of the most commonly used approaches, both dynamically and statically, for the study of the glass transition.

2.2 Mode Coupling Theory

In order to describe the dynamics of supercooled liquids and colloids, Bengtzelius et al. [6] and coworkers introduced simultaneously but independently the Mode Coupling Theory (MCT), which is a cornerstone in the glass literature. In this chapter we aim to present a general formulation of the theory, whose qualitative and quantitative predictions can be tested in experiments and computer simulations. Two main hypotheses deserve our attention, giving the basis on which MCT is built: (i) the structural properties of the glass former are rather similar to that of the high-temperature liquid; (ii) two separated time scales play an important role, a microscopic (fast) time scale associated with dynamical processes and a second one, accounting for relaxation processes and occurring on longer time scales. The MCT formalism makes use of these two hypotheses to derive a general equation of motions, after integrating out fast degrees of freedom.

2.2.1 Mori-Zwanzig Formalism

This Section is devoted to the explanation of the projection operator formalism according to Zwanzig's method [59]. After a rather broad overview of the theory, we will underline the main predictions and failures of the theory.

For the following description we consider a general N-particle system and an arbitrary phase-space function $f(p, q)$. The equation of motion of f can be written as:

$$\dot{f} = i\mathcal{L}f = \{H, f\} , \qquad (2.9)$$

where H is the Hamiltonian operator and $\mathcal{L}$ is the Liouville operator. The symbol $\{\cdot, \cdot\}$ denotes the Poisson brackets. By selecting a set of phase-space functions, $\{A_i\}$, where $i = 1, 2, \ldots N$, we can define a projection operator $\mathcal{P}$ such that $\mathcal{P}^2 = \mathcal{P}$:

$$\mathcal{P}f(t) = \sum_{i,j}^{N} A_i(0)\left[\langle A(0)|A(0)\rangle^{-1}\right]_{ij}\langle A_j(0)|f(t)\rangle . \tag{2.10}$$

The time evolution is given by $A_i(t) = \exp(i\mathcal{L}t)A_i(0)$. Now using the identity operator $\mathcal{P} + (1 - \mathcal{P})$, the evolution of the functions A_i reads:

$$\frac{dA_i(t)}{dt} = \exp(i\mathcal{L}t)[\mathcal{P} + (1 - \mathcal{P})]i\mathcal{L}A_i(t) = i\sum_j \Omega_{ij}A_j(t) + \exp(i\mathcal{L}t)(1 - \mathcal{P})i\mathcal{L}A_i(0),$$
$$\tag{2.11}$$

where the first term is a frequency matrix, i.e.

$$i\Omega_{ij} = \sum_k \left[\langle A(0)|A(0)\rangle^{-1}\right]_{ik}\langle A_k(0)|i\mathcal{L}A_j(0)\rangle , \tag{2.12}$$

and the second term can be rewritten by decomposing the evolution operator in a standard part and in a fluctuating force, $f_i(t)$,

$$\exp(i\mathcal{L}t)(1 - \mathcal{P})i\mathcal{L}A_i(0) = \int_0^t d\tau \exp[i\mathcal{L}(t - \tau)]i\mathcal{P}\mathcal{L}f_i(\tau) + f_i(t) , \quad (2.13)$$

$$f_i(t) = \exp[i(1 - \mathcal{P})\mathcal{L}t]i(1 - \mathcal{P})\mathcal{L}A_i(0) . \tag{2.14}$$

It is worth highlighting the role of the fluctuating force: its time evolution is driven by the complementary projection operator $\exp[i(1 - \mathcal{P})\mathcal{L}t]$, $1 - \mathcal{P}$ being the projector operator orthogonal to the space defined by $\mathcal{P}$. The presence of a term $(1 - \mathcal{P})$ yields an orthogonality condition between the fluctuating force and A_i. Physically, this is equivalent to removing the slow component—potentially provided by A_i— and propagate the force in the orthogonal space. In other words, the fluctuating force captures the remaining fast component of the force.

Using Eq. (2.10) and other well-known properties of the projector operator, we can write the equation of motion (2.11) as:

$$\frac{dA_i(t)}{dt} = \sum_{j=1}^{N}\left[i\Omega_{ij}A_j(t) - \int_0^t d\tau M_{ij}(\tau)A_j(t - \tau)\right] + f_i(t) . \tag{2.15}$$

Normally, we are interested in understanding how variables decorrelate in time. To quantify this effect, we shall define the time correlation function as $C_{ij}(t) = \langle A_i(0)|A_j(t)\rangle$. Because of the orthogonality condition $\langle A_i(0)|f_j(t)\rangle = 0$, the third term in (2.15) does not appear in the resulting expression for the correlation function [46]:

$$\frac{dC_{ij}(t)}{dt} = i \sum_{k=1}^{N} \Omega_{ik} C_{kj}(t) - \int_0^t d\tau \sum_{k=1}^{N} M_{ik}(\tau) C_{kj}(t - \tau) \,. \tag{2.16}$$

This relation depends on the memory function $M_{ij}(t)$, which in turn can be defined as:

$$M_{ij}(t) = \sum_k \left[\langle A(0)|A(0) \rangle^{-1} \right]_{ik} \langle f_k(0)|f_j(t) \rangle \,. \tag{2.17}$$

Equation (2.15) defines a *generalized Langevin equation* which can be solved by making use of Laplace transforms. However, the main difficulty comes from the complicated procedure used to determine the memory function. A way to tackle this problem is to identify A_i with the complete set of all possible slow variables and then to re-express the memory function as linear combinations of appropriate components of A_i. This approach defines the so-called mode-coupling equations. The most difficult task is then to extrapolate the dominant slow modes, whose number is in principle infinite.

2.2.2 Application of the Mori-Zwanzig Formalism to the Physics of Supercooled Liquids

By transposing this formalism to the context of supercooled liquids, the role of slow variables is played by the density fluctuations $\delta\rho(\vec{q}, t)$ where $\vec{q}$ is the wave vector [29, 46]. Therefore, we can write the following expressions for the density fluctuations and their correlation function $F(q, t)$, i.e. the *coherent intermediate scattering function*[6]:

$$\delta\rho(\vec{q}, t) = \sum_{j=1}^{N} \exp\left(i\vec{q} \cdot \vec{r}_j(t)\right), \quad F(q, t) = \frac{1}{N} \langle \delta\rho(\vec{q}, t)\delta\rho^*(\vec{q}, 0) \rangle \,. \tag{2.18}$$

The second expression, evaluated at $t = 0$, determines the *static structure factor* $S(q)$, which allows us to write the MCT-like equations for $\Phi(q, t) = F(q, t)/S(q)$:

$$\ddot{\Phi}(q, t) + \Omega^2(q)\Phi(q, t) + \int_0^t \left[M^{reg}(q, t - t') + \Omega^2(q)M(q, t - t') \right] \dot{\Phi}(q, t')dt' = 0 \,, \tag{2.19}$$

where $\Omega^2(q) = q^2 K_B T/(mS(q))$ is the squared frequency and m is the particle mass. The term $M^{reg}(q, t)$ represents instead the regular part of the memory function and governs the time dependence of $F(q, t)$ in the ballistic regime, that is for microscopic times. Conversely, the long-time behavior of $F(q, t)$ is governed by $M(q, t)$, which depends on the vertex $V^{(2)}$ and is bilinear in $\Phi(q, t)$:

---

[6]If we assume that space is isotropic, $F(q, t)$ depends on the wave-vector modulus only.

$$M(q,t) = \frac{1}{2(2\pi)^3} \int d\vec{k} \, V^{(2)}(q,k,|\vec{q}-\vec{k}|) \Phi(k,t) \Phi(|\vec{q}-\vec{k}|,t) \qquad (2.20)$$

$$V^{(2)}(q,k,|\vec{q}-\vec{k}|) = \frac{n}{q^2} S(q) S(k) S(|\vec{q}-\vec{k}|) \left(\frac{\vec{q}}{q} \left[\vec{k} c(k) + (\vec{q}-\vec{k}) c(|\vec{q}-\vec{k}|) \right] \right)^2 .$$
$$(2.21)$$

The other factor $c(k)$ in the vertex expression is the *direct correlation function*, which depends both on the density n and on the above-mentioned static structure factor, namely:

$$c(k) = n \left(1 - \frac{1}{S(k)} \right) . \qquad (2.22)$$

Equation (2.19) is akin to the expression for a damped harmonic oscillator with a time-dependent damping $M(q,t)$: the lower the temperature the greater the damping factor. It can be shown that the memory kernel plays the role of the variance of the random force acting on the density field: it includes contributions for all degrees of freedom, not only for the density field. It might affect the decay of $\Phi(q,t)$ and drive a non ergodicity transition. A fundamental role is also played by the vertex term since a large vertex contribution can increase the time scale at which the correlators decay to zero. This phenomenon is equivalent to saying that there exists a critical temperature at which the correlators no longer decay to zero, thus revealing a transition from an ergodic phase to a non-ergodic one, or, in other words, from the liquid to the solid disordered phase.

In principle, the correlation functions derived from the above set of equations can be compared with experimental data and computer simulations. However, finding an exact solution to this set of equations is very challenging. Various schematic models have been proposed to simplify the structure of the equations and to factorize the memory kernel [52, 53]. We do not want to enter into details of the multiplicity of these models. We only open a brief parenthesis on the formula (2.22), which depends on the static structure factor $S(q)$:

$$S(q) = 1 + 4\pi\rho \int_0^\infty dr \, r^2 \frac{\sin qr}{qr} (g(r) - 1) \qquad (2.23)$$

which can be experimentally measured via inelastic neutron scattering. In Eq. (2.23) ρ stands for the density, while $g(r)$ is the radial distribution function estimating the probability of finding a particle at a distance r from a tagged particle:

$$g(r) = \frac{1}{N} \frac{1}{4\pi r^2 \rho} \langle \sum_i^N \sum_{j \neq i}^N \delta(r - r_{ij}) \rangle \qquad (2.24)$$

$$r_{ij} = ||\vec{x}_i - \vec{x}_j|| . \qquad (2.25)$$

The average number of particles in a given shell $(r, r + dr)$ from a reference particle is $4\pi r^2 \rho g(r)$. However, the integration over r:

$$\int_0^\infty dr 4\pi r^2 \rho g(r) \tag{2.26}$$

is specific of the ensemble and, for instance, in the grand canonical ensemble it turns out to be proportional to the compressibility of the system χ_T [29].

The radial distribution $g(r)$ is zero at small distances[7] and normalized to one in the $r \to \infty$ limit, and has a nontrivial expression for a generic r. It displays a sequence of peaks that are informative of shells of neighbors around a given particle. In a crystal the peak distribution—the *Bragg-peak* distribution—is very sharp, while in the liquid phase, since there is no long-range order, there are just a few well-defined peaks that appear weaker and weaker at large distances.

However, the static structure factor alone is not really informative. From a static perspective, the particle arrangement in a supercooled liquid close to the glass transition temperature is pretty similar to that of a glass below T_g. Upon decreasing the temperature, the structure of the peaks changes only modestly.[8]

2.2.3 Dynamical Correlation Functions

Static observables do not provide any quantitative information deep into the supercooled liquid phase and close to the glass transition.[9] To be definite, the viscosity itself, which signals an abrupt change at the glass transition, is the integral over time of a *dynamical* correlator:

$$\eta = \int_0^\infty ds\, G(s) . \tag{2.27}$$

The same conclusions also hold for the diffusion coefficient, which is related to the integral of the velocity-velocity correlator. Hence, in the following we would like to analyze the typical behavior of dynamical correlation functions $C(t_1, t_2)$ for a generic observable $\phi_k(t)$, evaluated at time t for particle k:

$$C(t_1, t_2) = \frac{1}{N} \sum_{k=1}^{N} \langle \phi_k(t_1) \phi_k(t_2) \rangle . \tag{2.28}$$

---

[7]This vanishing behavior is a consequence of strongly repulsive forces at small distances.

[8]Note that both quantities, $S(q)$ and $g(r)$, are informative only far away from the glass transition and they are irrelevant to figure out whether the system is undergoing a phase transition.

[9]At first glance one might worry that the input of MCT equations, being purely static, should be unable to describe any more complex dynamical phenomenology. The right interpretation key relies on the nonlinear form of the vertex in Eqs. (2.20) and (2.21), which acts as a sort of feedback mechanism between statics and dynamics.

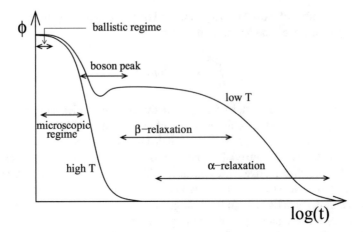

Fig. 2.3 Correlation function versus logarithmic time scale, for high temperatures (left side) and low temperatures (right branch) [35]. The exact nature of the so-called *boson peak* is very debated, but in many structural glasses it is supposed to be related to the existence of vibrational excitations characterized by a frequency that is one decade smaller than the typical vibrations of the system

At equilibrium the system is invariant under time-translation and, as a consequence, the correlation function depends only on the time difference. A convenient choice for $\phi_k(t)$ is the Fourier transform of the density fluctuations for the given particle k at fixed momentum. In this case $C(t)$ properly coincides with the intermediate scattering function $F_s(q, t)$. While in the high-temperature regime the correlation function $C(t)$ decays exponentially and is governed by a single time scale, both in the low-temperature region and very close to the glass transition temperature, the dynamic correlator exhibits a plateau and its decay is no longer purely exponential. This two-step relaxation is a qualitative signature of the approaching glass phase.

Let us consider more carefully the meaning of this plateau. We want to stress in particular two specific features: the time scale to reach the plateau and its duration. The former is not significantly affected by a temperature change, whereas the size of the plateau does, larger as temperature decreases. The first step relaxation to the plateau occurs on a quite short time scale, leading to the definition of β-relaxation regime (fast), whereas the second relaxation, so-called α-relaxation regime, takes place on a slower time scale.[10] The height of the plateau also holds information on the nature of the transition and is conventionally defined as the *non-ergodicity parameter*.

In Fig. 2.3 we show the schematic time dependence of the correlation function, wherein we can distinguish a first decay, which corresponds to the high-temperature relaxation in a normal liquid, and a second one, on the rightmost part of the plot, which exhibits a very slow relaxation trend. Let us consider first the high-temperature

---

[10]Several times earlier we referred to the relaxation time τ, which actually stands for the α-relaxation time.

region: at very short times that are shorter than the microscopic times, the behavior is purely ballistic. The intermediate region of the $\phi(t)$-plot, which—differently than the previous regime - is dominated by mutual particles interactions, is called *microscopic regime*. Finally, in the last region an exponential dependence can be highlighted, known as the *Debye relaxation*. A more complex situation occurs at low temperature: for short times there is again a ballistic regime, followed by the microscopic regime, but now the correlator is characterized by an intermediate plateau (β-relaxation). If one waits long enough, the correlator can decay to zero in the α-relaxation regime, whose start coincides with the end of the β-relaxation regime. By contrast to the behavior of normal liquids, the final decay of $\phi(t)$ can be approximated by the so-called *Kohlrausch-Williams-Watts function* [38, 57], which is a *stretched exponential*:

$$\phi(t) = A \exp\left[-(t/\tau)^{\beta}\right], \tag{2.29}$$

where A is an amplitude and the KWW-exponent β satisfies the condition $\beta \leq 1$. This anomalous behavior, which is not understood in terms of a Debye-relaxation process, is still an object of debate. It has generated two different scenarios, one supporting a heterogeneous point of view and another one, homogenous, which essentially reconnects the non-Debye relaxation to the presence of a generalized disorder.

A possible way to better understand the physical meaning of the plateau is to consider the mean-square displacement of a reference particle:

$$\langle r^2(t) \rangle \equiv \frac{1}{N} \sum_{i=1}^{N} \langle ||\vec{x}_i(t) - \vec{x}_i(0)||^2 \rangle . \tag{2.30}$$

This function describes the ballistic regime for short times and eventually gives rise to a diffusive regime. By plotting the mean square displacement of a supercooled liquid as a function of time on a logarithmic scale, one notices two regimes separated by a plateau that becomes more and more extended as temperature decreases. The emergence of a plateau is actually related to the vibrations of the particles inside *cages* formed by their neighboring particles (see Fig. 2.4). On the time scale corresponding to the plateau the particle is stuck in its cage and unable to move away. The α-relaxation time thus corresponds to the time needed for particles to escape their cages. On a much larger time scale, a diffusive behavior dominates the dynamics, which allows particles to move away from their cage. However, such description is only valid in real space, which can results in a misleading interpretation for some mean-field systems that do not have a space structure. In this case, a phase-space interpretation is preferred.

Fig. 2.4 The figure shows
the *cage effect* by zooming
into two internal particles
that are trapped by their
neighbors

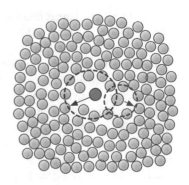

2.3 A Paradigmatic Example of Disordered System: The *p*-spin Spherical Model

To numerically and analytically handle MCT equations, reasonable approximations are needed, approximations which usually involve a non-perturbative scheme. Many critical and controversial aspects hover around this theory, leading to resulting expressions that might be un-controllable. However, Kirkpatrick and Thirumalai [34, 54], Kirkpatrick and Wolynes [33] proved that a MCT-like formalism can be safely applied to a general class of mean-field systems, namely the Potts glasses and the *p*-spin models. More precisely, spin auto-correlation functions in this type of systems verify the same equations of motion as those predicted by MCT. This very intriguing aspect will be a guiding principle of this thesis, which is focused on a strict connection between these two apparently different types of disordered systems: structural glasses and spin glasses. The class of models for which the mode-coupling equations are exact corresponds to mean-field statistical mechanics models. The dynamics can be exactly analyzed in the framework of the *p*-spin spherical model by writing the Langevin equation of motion [11], as follows:

$$\partial_t \sigma_i(t) = -\frac{\partial H}{\partial \sigma_i} - \mu(t)\sigma_i(t) + \eta_i(t) , \qquad (2.31)$$

where we introduced a Lagrange multiplier $\mu(t)$ to reproduce the spherical constraint, which prevents the field from departing in an unstable direction. The noise term η satisfies the following relations:

$$\langle \eta(t) \rangle = 0 \qquad \langle \eta(t)\eta(t') \rangle = 2T\delta(t - t') . \qquad (2.32)$$

The first term in the Langevin equation reads:

$$\frac{\partial H}{\partial \sigma_i} = -\frac{p}{p!} \sum_{kl} J_{ikl}\sigma_k\sigma_l , \qquad (2.33)$$

where we considered the $p = 3$ case, such that the couplings becomes J_{ikl}. Now we can write the generating functional which turns out to be composed of two terms:

$$i\hat{\sigma} \cdot \mathcal{L}_0 = \sum_k \int dt \; i\hat{\sigma}_k(t) \left[\partial_t \sigma_k(t) + \mu(t)\sigma_k(t) \right], \qquad (2.34)$$

$$i\hat{\sigma} \cdot \mathcal{L}_J = -\frac{ip}{p!} \int dt \sum_{ikl} J_{ikl} \hat{\sigma}_i(t)\sigma_k(t)\sigma_l(t). \qquad (2.35)$$

Because of the structure of the second term and the symmetry of couplings, we need to symmetrize the disordered part as well before averaging over J:

$$\overline{e^{i\hat{\sigma}\cdot\mathcal{L}_J}} = \int \prod_{i>k>l} dJ_{ikl} \exp\left\{ -\frac{1}{2p!} J_{ikl}^2 2N^{p-1} - J_{ikl} \int dt \left[i\hat{\sigma}_i\sigma_k\sigma_l + \sigma_i i\hat{\sigma}_k\sigma_l + \sigma_i\sigma_k i\hat{\sigma}_l \right] \right\}$$

$$= \exp\left\{ \int \frac{dt\,dt'}{4N^{p-1}} \left[p(i\hat{\sigma}\cdot i\hat{\sigma})(\sigma\cdot\sigma)^{p-1} + p(p-1)(i\hat{\sigma}\cdot\sigma)(\sigma\cdot i\hat{\sigma})(\sigma\cdot\sigma)^{p-2} \right] \right\}. \qquad (2.36)$$

The shorthand notation $\sigma \cdot \sigma$ identifies a two spin product evaluated respectively at time t and t'. We can now rewrite Eq. (2.36) in terms of a *dynamical overlap* [17, 51], namely the overlap between two given configurations at different times t and t':

$$\sigma(t) \cdot \sigma(t') \equiv \sum_k \sigma_k(t)\sigma_k(t')/N. \qquad (2.37)$$

In the following we summarize all the quantities of interest:

$$\begin{cases} Q_1(t,t') \equiv \langle i\hat{\sigma}(t) \cdot i\hat{\sigma}(t') \rangle = 0 \\ Q_2(t,t') \equiv \langle \sigma(t) \cdot \sigma(t') \rangle = C(t,t') \\ Q_3(t,t') \equiv \langle i\hat{\sigma}(t) \cdot \sigma(t') \rangle = R(t',t) \\ Q_4(t,t') \equiv \langle \sigma(t) \cdot i\hat{\sigma}(t') \rangle = R(t,t') \end{cases} \qquad (2.38)$$

where the first correlator must be zero, as argued in [51]. Indeed, $\langle \hat{\sigma}\hat{\sigma} \rangle = 0$ is a self-consistent solution at any temperature. The presence of such a finite correlator would generate closed loops responsible for a spontaneous symmetry breaking and for a violation of the partition function normalization.

At this point, using the δ-function representation in terms of an exponential function and setting to zero the derivatives with respect to all the Q's, we can write mutual relations between them. Considering the simplest case with a scalar degree of freedom σ, the effective Langevin equation can be written as:

$$\partial_t \sigma(t) = -\mu(t)\sigma(t) + \frac{1}{2}p(p-1)\int dt'' R(t,t'')C(t,t'')^{p-2}\sigma(t'') + \xi(t), \quad (2.39)$$

where $\xi(t)$ is no longer a delta-correlated white noise, but:

$$\langle \xi(t)\xi(t') \rangle = 2T\delta(t - t') + \frac{p}{2}C(t, t')^{p-1} . \tag{2.40}$$

In other words, there is a non-local kernel that couples the selected time t with all the previous times $t'' < t$. Before pursuing the discussion, we introduce a useful formal relation for the response function:

$$R(t, t') = \left\langle \frac{\delta\sigma(t)}{\delta\xi(t')} \right\rangle . \tag{2.41}$$

Then, differentiating the effective Langevin equation (2.39), we obtain the following expressions, which are at the core of the MCT derivation:

$$\frac{\partial R(t_1, t_2)}{\partial t_1} = \left\langle \frac{\delta\dot{\sigma}(t_1)}{\delta\xi(t_2)} \right\rangle = -\mu(t_1)R(t_1, t_2) + \frac{1}{2}p(p-1)\cdot$$

$$\int_{t_2}^{t_1} dt'' R(t_1, t'')C^{p-2}(t_1, t'')R(t'', t_2) + \delta(t_1, t_2) , \tag{2.42}$$

$$\frac{\partial C(t_1, t_2)}{\partial t_1} = \frac{\partial}{\partial t_1}\langle \sigma(t_1)\sigma(t_2) \rangle = -\mu(t_1)C(t_1, t_2) + \frac{1}{2}p(p-1)\cdot$$

$$\int_{-\infty}^{t_1} dt'' R(t_1, t'')C^{p-2}(t_1, t'')C(t'', t_2) + \langle \xi(t_1)\sigma(t_2) \rangle , \tag{2.43}$$

where the last term in brackets can be calculated using [11]:

$$\langle \xi(t_1)\sigma(t_2) \rangle = \int dt'' D(t_1, t'')R(t_2, t'') = 2T R(t_2, t_1) + \frac{p}{2}\int_{-\infty}^{t_2} dt'' R(t_2, t'')C^{p-1}(t_1, t'') . \tag{2.44}$$

This formalism also gives the tools to derive the expression for the Lagrange multiplier $\mu(t)$:

$$\mu(t_1) = \frac{1}{2}p^2 \int_{-\infty}^{t_1} dt'' R(t_1, t'')C^{p-1}(t_1, t'') + T . \tag{2.45}$$

2.3.1 Connection with the Static Replica Computation

In order to see the connection we will now analyze the same p-spin spherical model in a static approach, where the Hamiltonian of N interacting spins reads:

$$H[\vec{\sigma}] = -\sum_{i_1 < ... < i_p}^{1,N} J_{i_1,...,i_p}\sigma_{i_1}...\sigma_{i_p} , \tag{2.46}$$

and the spin variables satisfy the spherical constraint $\vec{\sigma} \cdot \vec{\sigma} = \sum_{i=1}^{N} \sigma_i^2 = N$. The couplings $J_{i_1,\dots i_p}$ are independent random variable extracted from a Gaussian distribution with zero mean and variance $\overline{J^2} = \frac{p!}{2N^{p-1}}$. This choice of the couplings guarantees the correct extensive, $O(N)$, dimensional scaling of the Hamiltonian. While the partition function of the model, Z_J, depends on the realization of the quenched disorder J, the free energy is a *self-averaging* quantity [18], independent of the particular choice of disorder in the thermodynamic limit.

The replicated partition function, which is constrained to have the spins on a N-dimensional spherical manifold, can then be rewritten as:

$$\overline{Z^n} = \int \mathcal{D}\vec{\sigma}\, e^{-\frac{\beta\lambda}{2}\sum_{a=1}^{n}(\sigma^a \cdot \sigma^a - N)}\, \overline{e^{-\beta \sum_{a=1}^{n} H[\sigma^a]}} =$$

$$= \int \mathcal{D}\vec{\sigma}\, e^{-\frac{\beta\lambda}{2}\sum_{a=1}^{n}(\sigma^a \cdot \sigma^a - N)} \prod_{i_1 < \dots < i_p}^{1,N} \int \mathcal{D}J_{i_1,\dots,i_p} e^{-J_{i_1\dots i_p}^2 \frac{N^{p-1}}{p!} + \beta J_{i_1,\dots,i_p} \sum_{a=1}^{n} \sigma_{i_1}^a \dots \sigma_{i_p}^a}$$

$$\tag{2.47}$$

and then, performing the Gaussian integrals:

$$\overline{Z^n} = \int \mathcal{D}\vec{\sigma}\, e^{-\frac{\beta\lambda}{2}\sum_{a=1}^{n}(\sigma^a \cdot \sigma^a - N)} \prod_{i_1 < \dots < i_p}^{1,N} \exp\left[\frac{\beta^2 p!}{4N^{p-1}} \sum_{a,b}^{1,n} \sigma_{i_1}^a \sigma_{i_1}^b \dots \sigma_{i_p}^a \sigma_{i_p}^b \right] =$$

$$= \int \mathcal{D}\vec{\sigma} \exp\left[-\frac{N\beta\lambda}{2} \sum_{a=1}^{n}(Q_{aa} - 1) + \frac{\beta^2 N}{4} \sum_{a,b}^{1,n}(Q_{ab})^p \right],$$

$$\tag{2.48}$$

where we defined the overlap matrix as:

$$Q_{ab} = \frac{1}{N} \sum_{i=1}^{N} \sigma_i^a \sigma_i^b .$$

$$\tag{2.49}$$

Introducing a factor 1 and using the integral representation of the δ function,

$$\int \prod_{a,b}^{1,n} dQ_{ab} \int \prod_{a,b}^{1,n} d\mu_{ab} e^{\mu_{ab}(NQ_{ab} - \sum_{i=1}^{N} \sigma_i^a \sigma_i^b)} = 1 ,$$

$$\tag{2.50}$$

we can rewrite the replicated partition function as:

$$\overline{Z^n} = \int d\hat{Q} d\hat{\mu} e^{-NS(\hat{Q},\hat{\mu})} ,$$

$$\tag{2.51}$$

where the action $S(\hat{Q}, \hat{\mu})$ reads:

$$S(\hat{Q}, \hat{\mu}) = \sum_{a,b}^{1,n} \left(-\frac{\beta^2}{4} Q_{ab}^p - \mu_{ab} Q_{ab} \right) + \frac{1}{2} \ln \det(2\hat{\mu}) . \tag{2.52}$$

Note that the multiplier μ_{ab} also includes the case $a = b$, which imposes the spherical constraint. Thanks to the explicit dependence on N, we can perform a saddle-point computation, which is exact in the $N \to \infty$ limit. The free energy is then properly defined by a two-limit procedure:

$$- \beta F = \lim_{N \to \infty} \lim_{n \to 0} \frac{1}{nN} \log \int d\hat{Q} d\hat{\mu} e^{-NS(\hat{Q},\hat{\mu})} \tag{2.53}$$

where in principle the two limits cannot be generally inverted. In this case the only thing to do is to consider first $N \to \infty$, solve the integral exactly and then send $n \to 0$. This is a subtle point that should be taken carefully into account for any other model.

2.4   Fluctuation-Dissipation Theorem and the Dynamical Temperature

When looking at an equilibrium system, the fluctuation-dissipation theorem and time-translation invariance can be safely applied, which give rise to expressions for the correlation function $C(t, t')$ and the response function $G(t, t')$ no longer dependent on two times, but only on their difference $\tau = t - t'$. Through the fluctuation-dissipation theorem a direct link between the two can further be established:

$$G(\tau) = -\frac{1}{T} \frac{dC(\tau)}{d\tau} . \tag{2.54}$$

After some algebraic manipulations [11], the coupled equations derived in (2.43) for the p-spin can then be reduced to:

$$\dot{C}(\tau) = -TC(\tau) - \frac{p}{2T} \int_0^\tau dt'' C^{p-1}(\tau - t'') \dot{C}(t'') , \tag{2.55}$$

where we imposed that $C(\infty) = 0$, i.e. no ergodicity breaking occurs. Let us consider what the two terms in Eq. (2.55) represent: the first term on the r.h.s is due to the disorder-independent part of the Langevin equation and more exactly to the spherical constraint, while the second one, a memory term, is obtained by integrating over J. By using the fact that the average correlation decreases with time (i.e. $\dot{C}(\tau) \le 0$), we obtain an upper bound:

$$C^{p-2}(\tau) (1 - C(\tau)) \le \frac{2T^2}{p} \tag{2.56}$$

that can be solved graphically. If we relabel the l.h.s of (2.55) as $f(C)$, we note that for $\tau = 0$, $C = 1$ and $f(1) = 0$, whereas for $\tau \to \infty$, $C(\infty) = 0$ and $f(0) = 0$. As a consequence, the function $f(C)$ has a maximum in $q_d = \frac{p-2}{p-1}$.

The inequality in Eq. (2.56) is surely satisfied in the high-temperature phase, namely in the ergodic phase, where $f(q_d) \ll 2T^2/p$. Upon decreasing the temperature, the difference between the l.h.s and the r.h.s reduces: this effect is responsible for the emergence of a plateau where $C(\tau) \sim q_d$.

The temperature for which Eq. (2.56) is exactly satisfied, i.e. $2T_d^2/p = f(q_d)$, is known as the *dynamical temperature*:

$$T_d = \sqrt{\frac{p(p-2)^{p-2}}{2(p-1)^{p-1}}} \,. \tag{2.57}$$

At this temperature a kinetic arrest occurs[11] because the system remains trapped at a plateau with $\dot{C} = 0$ and thus undergoes a transition from the paramagnetic phase to a non-ergodic phase with a finite self-overlap q_d. The transition can be of two types: a continuous transition with a static temperature T_s equal to the dynamical temperature T_d and a discontinuous transition for which $T_s < T_d$. In the replica language this means that the system exhibits a one-step replica symmetry breaking, i.e. a first-order transition in terms of the order parameter, but a second-order transition in terms of the underlying thermodynamics.

In summary, the mode-coupling theory predicts a glass transition at a finite temperature T_c, also referred to as T_d. It also predicts a power-law behavior of transport factors, such as the relaxation time τ, regulated by a universal exponent γ:

$$\tau(q, T) \sim (T - T_c)^{-\gamma} \,. \tag{2.58}$$

The inverse of the diffusion constant follows an analogous scaling:

$$D^{-1}(T) \approx \tau(q, T) \,. \tag{2.59}$$

Below the critical temperature T_c, an exponential number of states emerges, giving rise to a very complex landscape. In principle, the partition function can be calculated by summing over all these states, each weighted by their corresponding Boltzmann factor. The key point is that below T_c there is a competitive mechanism

---

[11]We need to distinguish two cases: for $T > T_d$ our analysis is correct and sufficient, supporting the idea that the correlation function is characterized by two exponents given in Eq. (2.63), while for $T < T_d$ time-translation invariance does not hold anymore, giving rise to a more complex dynamics influenced by *aging effects*. The underlying physics background of this second case is rather different. For *aging* we mean that the properties of the system strongly depend on its age and history, namely on the waiting time between the sample preparation and the begin of measurements. Time-translation invariance does not hold anymore and out-of-equilibrium features are observed. This scenario characterizes several materials, from polymers to molecular glasses, to spin glasses as well.

between single-state free energies, which tend to minimize the global free energy, and configurational entropy, which favors states with the highest free energy.

As depicted in Fig. 2.3, we can distinguish two different regimes: a so-called β regime, when the solution $\Phi(t)$ approaches the plateau, and a second one known as α regime, which is responsible for the departure from it. MCT makes interesting predictions for several observables, in particular for the behavior of arbitrary correlators near their plateau:

$$\Phi(t) = \Phi_{EA} + h_x G(t), \tag{2.60}$$

where Φ_{EA} is the plateau height (also known as *Edwards-Anderson parameter*), which measures the fraction of memory of the initial state that still persists. The other two factors are respectively an amplitude and a universal function of time.[12] According to the theory, $G(t)$ presents a peculiar dependence on temperature given by:

$$G(t) = \sqrt{|\sigma|} g_{\pm}(t/t_\sigma), \tag{2.61}$$

where σ is the rescaled temperature $(T_c - T)/T_c$ and $t_\sigma = t_0 |\sigma|^{1/2a}$, depending on a microscopic time scale t_0. The positive or negative sign that characterizes the function g refers to $T < T_c$ and $T > T_c$ respectively, while its argument, the time scale t_σ, has a power-law divergence regulated by σ. Despite the fact that an exact solution for $g_{\pm}$ cannot be derived analytically in a broad interval, its asymptotic form can be deduced close to the plateau, i.e.:

$$
\begin{aligned}
g(t) &= t^{-a} & t \ll 1 &\leftrightarrow & t_0 \ll t \ll t_\sigma \\
g(t) &= t^{b} & t \gg 1 &\leftrightarrow & t_\sigma \ll t \ll \tau .
\end{aligned}
\tag{2.62}
$$

The first functional form gives rise to a *critical decay*, while the latter is known as the *von Schweidler law* [28]. The two exponents, a and b, which identifies the separation between the two relaxation scales, are mutually related:

$$\frac{\Gamma(1-a)^2}{\Gamma(1-2a)} = \frac{\Gamma(1+b)^2}{\Gamma(1+2b)} = \lambda, \tag{2.63}$$

where $\Gamma(x)$ is the ordinary Γ-function. The two values a, b are not arbitrary. They must satisfy the following conditions: $0 \le a < 1/2$, $0 \le b \le 1$ in agreement with empirical data. Moreover, the exponent γ, which appears in Eq. (2.58), is given by:

$$\gamma = \frac{1}{2a} + \frac{1}{2b} . \tag{2.64}$$

It is worth remarking that MCT predicts a sharp glass transition at an incorrect temperature T_c. It predicts a dynamical arrest when the system is still ergodic and

---

[12]In writing this expression we have assumed the *factorization property* i.e. that the difference between the correlator and its plateau value is simply a function of two independent contributions.

not yet an amorphous solid.[13] This is in agreement with the argument suggesting a strict connection between a mode coupling approach and a mean-field theory [2], in which the free energy barriers are greater than the finite-dimensional case and the transition is reached before the real case.[14] Moreover, MCT predicts a power-law divergence of the transport coefficients that is not accurate over a broad temperature range. An incorrect analysis of non-Gaussian parameters and the lack of a breakdown of the Stokes-Einstein relation[15] are other weak points of this theoretical approach. Despite these inaccuracies, it provides remarkable scaling properties in the β-relaxation regime and useful time-temperature relations in the α-relaxation regime. It can thus be considered an important framework to get analytical predictions in models with quenched disorder for which schematic MCT-like equations are exact.

2.5 Dynamical Facilitation and Kinetically Constrained Models

In this zoology of glassy phenomenologies, another interesting approach is based on the notion of *dynamical facilitation* [15]. In this framework the dynamics is substantially due to facilitation effects, i.e. defects that trigger the relaxation of nearest molecules. Typically motility in a viscous liquid cannot spontaneously emerge (or conversely be eliminated) if there are no mobile particles around. This mechanism dominates the physics near T_g even though unsuitable to capture the microscopic origin of the dynamical slowing down. According to the dynamical facilitation mechanism, several lattice models have been proposed, known as *kinetically constrained models* (KCM). Their thermodynamics is usually not interesting, described by a non-interacting lattice gas Hamiltonian. What actually plays a relevant role is the dynamics. If a kinetic constraint is satisfied, the system evolves allowing the passage from one configuration to another. Its dynamics follows a standard Markovian process determining a sequence of particle jumps. We refer here to local constraints, that nevertheless cause a collective dynamical behavior. The underlying mechanism is a dynamical frustration, namely in the high-density or low-temperature regime a competition between few excitations or defects and the necessity to facilitate local motion. Indeed both temperature and density might change the effective connectivity on the lattice. Note that this frustrating mechanism generates a cooperative relaxation

---

[13]Note, however, that in real glasses nucleation processes happen, preventing the emergence of a transition at T_d. It would be replaced by a crossover regime within which the relaxation times increase very fast at low temperatures.

[14]The incorrect transition predicted in MCT is supposed to be due to extreme simplifications that do not take into account activated barrier crossings. This is why below the critical temperature, T_c, the theory is completely unsuitable.

[15]Over the years many efforts have been done to correctly describe the mechanism that leads to such a breakdown, once critical fluctuations are taken into account: a field theory argument is for instance proposed in [8].

process, which is very different from the RFOT mechanism we shall explain in the next Section (even if also based on the concept of cooperativeness).

KCMs can be classified into *conservative* and *non-conservative* models, depending on whether the total number of occupied (or empty) sites is taken fixed or changes in time. The second class of models defines a sort of coarse-graining, as each lattice site corresponds to a small region with a certain number of particles. Depending on whether their density is above or below a given threshold one can populate the state and then have a local rearrangement. Independently from their classification, the rates satisfy a detailed-balance relation with a Boltzmann distribution factorized over the sites. Then the only static interaction is due to the hard-core constraint according to which $n_i \in \{0, 1\}$.

In the following we shall give some relevant definitions by focusing on two very well-known examples, respectively belonging to the conservative and the non-conservative class. Let us start describing the so-called Kob-Andersen model (KA) [36, 37], whose Hamiltonian reads:

$$H[\{n_i\}] = 0 , \qquad n_i = 0, 1 \tag{2.65}$$

allowing a particle to jump in a given site only if empty. Moreover, the sites occupied before and after the move must have a number of neighbors fewer than a tunable parameter m. In the original formulation, $m = 3$ with particles located on a cubic lattice. Several generalizations can be done for example on a hyper-cubic lattice in d dimensions, with $2 \leq m \leq d$. As mentioned before, the crucial difference with the usual lattice glass models lies in the fact that here the geometrical constraints act as kinetic rules, with no influence played by the thermodynamics or the interactions between the sites, except for the hard-core constraint. These models seem to properly capture the idea of a *cage effect*, i.e. if the neighbor shell is particularly dense the particle cannot diffuse.

Several variations have been proposed, for instance focusing on holes rather than particles. A well-known model was proposed by Fredrickson and Andersen (FA) [24] in 1984, providing an alternative approach to study the glass transition by means of simple kinetic considerations. Its Hamiltonian reads:

$$H[\{n_i\}] = J \sum_{i=1}^{N} n_i , \qquad n_i = 0, 1 \tag{2.66}$$

where one can assume to have either non-interacting spins or defects as variables. J stands for a sort of activation energy determining the mobility in the system. Its value determines the average defect concentration, which is exponentially small in the low-temperature regime, i.e. $\langle n_i \rangle \approx \exp(-J/T)$. The transition rule is given by a Glauber dynamics with the condition that a particle can jump only if the chosen site is surrounded by a certain number of defects fixed a priori.

One can thus define two different classes, one based on *non-cooperative* models (e.g. taking the number of nearby defects $k = 1$ in the aforementioned model) which display an Arrhenius behavior, and the other based on *cooperative* models, which are characterized by super-Arrhenius dependance of the relaxation time (e.g. considering $k > 1$ or again with $k = 1$ imposing a directional bias). A similar definition also holds for conservative KCMs.

Both the FA and the KA models in their simplest version on the Bethe lattice display an ergodicity breaking transition, while their modified version FA-m and KA-m for any choice of m and any dimension d on a hyper-cubic lattice do not have such transition. The main reason for which these models have grabbed so much attention is without any doubt due to their simple definition, which nevertheless captures several interesting phenomena. These include the super-Arrhenius or Arrhenius slowing down (depending on the specific model), the natural emergence of dynamical heterogeneities, the possibility to introduce several invariant measures on them and last but not least their aging phenomena. Moreover, these models turn out to be very beneficial to study the physics of hard-sphere systems and jamming in general [48, 50, 56], which will be the central theme of Chap. 3.

As every theory that has pros and cons, the facilitation mechanism cannot provide a full explanation of all thermodynamic quantities, characterized by a non-trivial phenomenology (e.g. the entropy crisis), and in the same way cannot do a good job in terms of static correlation functions. Facilitation turns out to be a rather strong assumption, affecting the distribution of defects and vacancies. In other words, since this mechanism can neither create nor destroy mobile particles/defects, any spontaneous motion is prevented at all. On the other hand, if one intended to remove the mobility conservation hypothesis, KCMs as well as their glassy phenomenology would become trivial [7].

There is one more aspect that makes this approach prone to criticism: indeed, none of these models can be derived from microscopic first-principle calculations and their dynamical rules are just imposed from outside. Then, we aim to discuss now an approach that heads in the opposite direction.

2.6 A Static Approach: Random First Order Transition Theory

At the beginning of this chapter, we mentioned the configurational entropy crisis as proposed by Kauzmann. The basic underlying idea is that the system ends up in a single stable state wherein it is unable to reach another state, even in an infinite amount of time. At least in fragile glass formers, the mechanism leading to this phase transition and regulating potential energy barriers, whose height increase upon decreasing temperature, turns out to be particularly interesting. The physical reason behind this growth can be understood in terms of cooperative rearrangements of particles in real space. This process, in turn, is reminiscent of the concept of an increasing correlation length and an exponentially increasing relaxation time. We are essentially interpreting the super-Arrhenius behavior in fragile glasses as the consequence of larger rearranging regions, which become more and more correlated at lower temperatures. A natural question that arises is whether a correlation length exists and it can be detected in a simple way.

2.6.1 The Adam-Gibbs-Di Marzio Theory

The first attempt to connect an increasing correlation length to a vanishing configurational entropy was proposed in a completely different context by Flory, who was interested in computing the number of possible configurations in a polymer [19]. In the framework of glass-forming liquids, Adam, Gibbs and Di Marzio were the first to seriously think about the connection between T_K and a more complex phenomenology. Below T_K ergodicity breaks down and a new phase of matter emerges, *a fourth phase of matter* as defined by the authors [27]. This phenomenon has been more formally explained in [1] in terms of free-energy barriers and a more accurate difference between *states* and *configurations*. The starting point of the theory we present here is the introduction of the so-called Cooperative Rearranging Regions (CRR), defined as the smallest regions that can be rearranged independently from their surrounding. The number of locally stable states to which each CRR can belong, indicated as Ω, is constant, affected neither by temperature changes or by other specific quantities in the system. Using the hypothesis of a weak interaction between different CRRs, one can write the total number of states in which the system can be as:

$$\mathcal{N} = \Omega^{N/n} \tag{2.67}$$

where N stands for the total number of particles in the system and n is the number of particles in a typical CRR. From this, the total configurational entropy reads:

$$S_c(T) = \frac{1}{N} \log \mathcal{N} = \frac{1}{n(T)} \log \Omega . \tag{2.68}$$

Focusing the attention on the size of a typical CRR, we get:

$$n(T) = \frac{\log \Omega}{S_c(T)} , \tag{2.69}$$

immediately suggesting that a diverging length scale corresponds to a vanishing configurational entropy. Note that, although n identifies the number of particles in a typical CRR, it is also related to its linear size ξ as $n \sim \xi^d$. This is one possible choice, susceptible to modifications and improvements, as we will better clarify in the following.[16]

---

[16]Originally, Adam and Gibbs assumed compact CRRs but this is not essential. A different scaling can also be considered that will give rise, in turn, to a different exponent in the relaxation time and the critical correlation length.

Assuming that the free-energy barrier to overcome in order to have a cooperative rearrangement is proportional to the size of the CRR, the relaxation time then reads:

$$\tau_R = \tau_0 \exp\left(\beta \Delta F\right) = \tau_0 \exp\left(\frac{B}{T S_c(T)}\right) \tag{2.70}$$

where, as explained for the Arrhenius and super-Arrhenius laws, τ_0 is just a normalizing prefactor, while B accounts for constant terms. Deriving an exact relation for τ_R is beyond our control as the behavior of the configurational entropy is unknown. However, we can make use of the relation linking the entropy to the specific heat in a thermodynamic system and we can assume that the latter does not depend on temperature to obtain:

$$S_c(T) - S_c(T_K) = \int_{T_K}^{T} dt \frac{\Delta c_p}{T} \,, \tag{2.71}$$

where Δc_p is the difference between the specific heat of the liquid and that of the crystal. The expression above leads to:

$$S_c(T) = \Delta c_p \log(T/T_K) \sim \Delta c_p \frac{T - T_K}{T_K} \,, \tag{2.72}$$

after expanding the logarithm for $T \sim T_K$. Hence, by plugging this expression in Eq. (2.70) and assuming that the system is sufficiently far from $T = 0$, we recover the same trend for the relaxation time as predicted by the VFT law. Albeit based on simple arguments, this theory has achieved resounding success over the years and has provided a simple linear expansion in agreement with empirical data. However, some debated points will be addressed later, in the framework of the mosaic theory. Two main hypotheses in particular have been long critically examined: (i) considering a constant number of metastable states, despite the increasing size of the region; (ii) describing the growth of the barrier proportionally to the volume of the CRR. Concerning the first point, from a meticulous analysis of the data provided by Adams and Gibbs, Johari [32] claimed that for many materials (including glucose and glycerol) the typical size of a CRR turns out to be less than one. Even relaxing the hypothesis on Ω and assuming a more reasonable value for it, for about thirty materials it can be seen that the typical size remains $<O(10)$. The second object of debate is that, since the barrier grows like the volume of a region of linear size ξ, in order to have a cooperative rearrangement, a finite fraction of the total number of particles has to be taken into account. However, this is unfeasible for large ξ [41].

2.6.2   The Mosaic Theory

Some of these critical issues have been clarified within the mosaic theory, also known as Random First Order Transition Theory (RFOT).[17] It can be regarded as an effort to go beyond mean-field models and describe via simple nucleation arguments the emergence of a complex free-energy landscape with an enormous number of trapping minima. Indeed, in finite dimension the notion of metastable state does not make proper sense. The presence of a surface tension is responsible in turn for the presence of many amorphous states, which are all statistically equivalent, at least locally.

Let us suppose that initially the system is in a state α. Thanks to thermal fluctuations it might evolve into another state β in a finite time, in agreement with the previous argument based on cooperative rearrangements. This α-β linking mechanism is not exonerated from paying a free-energy price due to the formation of an interface between this emerging region and the rest of the system. To formalize this argument let us write the droplet activation free energy $\Delta F(\xi)$, consisting of a driving force and of an energy cost proportional to a generalized surface tension $\Upsilon(T)$:

$$\Delta F(\xi) \sim -TS_c(T)\xi^d + \Upsilon(T)\xi^\theta , \tag{2.73}$$

where θ must be at most $d-1$ to reproduce a surface effect in d dimensions. As in the standard nucleation picture, for small enough ξ, the leading contribution is the cost term that prevents the formation of a droplet. Conversely, when ξ increases beyond the critical value:

$$\xi_c = \left(\frac{\Upsilon(T)}{TS_c(T)} \right)^{\frac{1}{d-\theta}} , \tag{2.74}$$

the entropic term plays the leading role, hence the nucleation of the liquid into the glassy state is favored. The relaxation time associated with this critical length, as reported in Eq. (2.74), is:

$$\tau(\xi_c) = \tau_0 \exp \left\{ \left(\frac{\Upsilon(T)}{TS_c(T)} \right)^{\frac{\theta}{d-\theta}} \right\} . \tag{2.75}$$

This critical correlation length ξ_c has relevant physical implications especially in terms of metastable states. To clarify this point, let us suppose that we can prepare the system at $T=0$ in a global energy minimum. Then we increase temperature with the aim of turning this global minimum into a global metastable state. Note that this is just a naïve perspective because thermal fluctuations hinder the presence of metastable states. Any global configuration is unstable against the formation of droplets of size comparable to ξ_c. After a while the system thus ends up in a het-

---

[17]This peculiar definition denotes a second-order phase transition in the usual thermodynamic sense, according to Ehrenfest's classification, but accompanied by a discontinuous jump in the order parameter. Models displaying a RFOT are actually described in the replica formalism by a one-step replica symmetry breaking.

erogenous, *mosaic*-like structure. Several local states emerge in a continuous process of rearrangements occurring on a typical scale ξ_c. A really interesting *Gedankenexperiment* has been proposed in [10] in an effort to clarify that only below a given length a description in terms of metastable states is reasonable. It is worth mentioning that even though generated from the same conceptual hypotheses, quantitatively the two aforementioned approaches, the Adam-Gibbs-Di Marzio and the mosaic theories, make different predictions. Concerning the correlation length at which the transition takes places, the mosaic theory leads to $\xi_c \sim (1/S_c)^{1/d-\theta}$, whereas AGDM to $\xi_c \sim (1/S_c)^{1/d}$. As the exponent $\theta \leq d-1$, an increase in the correlation length would be sharper within the mosaic theory than in the AGDM framework. This is not the only difference between the two approaches. Their evaluation of the typical scaling of the energy barrier only coincides under specific assumptions. The resulting expression for the relaxation time predicted within the mosaic picture remains implicit, strongly dependent on the exponent θ. To determine the exponent θ, Kirkpatrick et al. performed a renormalization group approach based on the following assumptions[18]:

- the droplet should be large enough to be considered thermodynamically independent;
- an ideal glass transition occurs at T_K;
- critical scalings should be considered near T_K.

As a consequence, one should be able to connect the correlation length ξ of the emerging phase to a reduced temperature $\tilde{t}^{-\nu}$ where $\tilde{t} = \frac{T-T_K}{T_K}$ and ν is the exponent for the correlation length, as reported in the critical phenomena literature [30]. In the theory of critical phenomena, indeed, near the critical temperature the correlation length is expected to have a power-law scaling with an exponent ν that is equal to 1/2 in mean-field approximation. However, we want to consider here a general case without making any assumption on the exponent.

Therefore, inserting the scaling relation $\xi \sim \tilde{t}^{-\nu}$ in the droplet free-energy expression:

$$\Delta F(\xi) \sim -T S_c \xi^d + \Upsilon \xi^\theta , \qquad (2.76)$$

one gets:

$$\Delta F(\tilde{t}) = -\tilde{t}^{1-d\nu} + \tilde{t}^{-\theta\nu} . \qquad (2.77)$$

A cooperative rearrangement of particles needs to be energetically favored and this is possible only if the first term—the driving force—scales faster than the second, i.e. if $1 - d\nu \leq -\theta\nu$. This yields:

$$\theta \leq \frac{d\nu - 1}{\nu} . \qquad (2.78)$$

We need now to link the exponent ν to the fluctuation relation for T, where k_B is the Boltzmann factor and C the specific heat, namely:

---

[18]For more details we refer the reader to [41].

$$(\delta T)^2 = \frac{k_B T^2}{nC} .$$ (2.79)

Making use again of the theory of critical phenomena, according to which $C \sim \tilde{t}^{-\alpha}$, and assuming that one is sufficiently close to T_k, we can expand the configurational entropy and then get $\delta T \leq T - T_K = \tilde{t} T_K$. Collecting all these results, we recover:

$$\alpha + d\nu \geq 2 ,$$ (2.80)

where for $\alpha + d\nu = 2$ is the usual hyper-scaling relation [30], in which d appears explicitly. Under further hypotheses to describe the discontinuity in the specific heat at the transition, one gets $\theta \sim d/2 \sim 1/\nu$. This result immediately allows us to understand the correspondence between this approach and the VFT law. Albeit in agreement with the predictions of one of the most used fitting laws, there is no general consensus on this argument. The main hypotheses that serve as the building block for renormalization group formalism, have been largely debated. None of the above assumptions has been proven. Indeed, near T_K the typical size of a CRR is unknown because the material is already vitrified at T_g. At T_g the estimated correlation length is too small to consider the independence of domains as a reasonable assumption.

Most recently the mosaic theory has been rephrased in a different way in order to connect the mosaic length ξ_c to a point-to-set correlation length [9, 20]. Thanks to these works, new prospects for numerical simulations have opened up.

2.7   Complexity in Mean-Field Systems

In the previous Section we mentioned the RFOT theory as a good starting point to better investigate glass formers. It constitutes, together with the method that we are going to present now, one of the most popular static approaches. Their cornerstone is the concept of metastability.

Let us start our discussion by saying that the equilibrium probability distribution can be always decomposed into a set of *pure states* α, which in the case of fully connected systems are completely determined by a set of local average observables (e.g. local magnetizations in magnetic systems, average particle positions in structural glasses, etc.). By contrast, in finite-dimensional systems pure states are only defined by taking the thermodynamic limit and fixing certain boundary conditions [25]. In mean-field-like systems, such as the p-spin model, we can nevertheless adopt a simplified description in which the total partition function can be thought as the contribution of all these pure states. In the following pages we want to determine a general potential as a function of a local order parameter, whose minima are in correspondence with the Thouless-Anderson-Palmer (TAP) states of the system [14, 43, 55, 58]. In this framework, below a certain temperature T_{TAP}, shown in Fig. 2.5, the number of these minima is exponential in the size of the system and these minima can be grouped according to their intensive free energy f_α. The partition can thus be

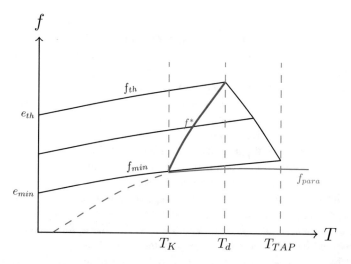

Fig. 2.5 Schematic plot of the TAP free energy density as a function of temperature. The blue line corresponds to the states with free energy f^*, which dominate the partition function over the interval (T_K, T_d). The complexity vanishes as the free energy approaches f_{min} at T_K, and above the maximum value f_{th}. Between the dynamical and the Kauzmann temperatures the red full line corresponds to the equilibrium free energy $f^* - T\Sigma(f^*)$. This figure refers to the spherical p-spin model which has a negative entropy (in principle not forbidden since the degrees of freedom are continuous variables) and a positive free energy, which is an increasing function of temperature. The situation is inverted in discrete models but the qualitative plot above remains valid

written as:

$$
\begin{aligned}
Z &= e^{-\beta N F} \sim \sum_\alpha e^{-\beta N f_\alpha} = \int df \sum_\alpha \delta(f - f_\alpha) e^{-\beta N f} \\
&= \int df \, (f) \, e^{-\beta N f} = \int_{f_{min}}^{f_{th}} df \, e^{N[\Sigma(f,T) - \beta f]} \sim e^{N[\Sigma(f^*,T) - \beta f^*]}
\end{aligned}
\tag{2.81}
$$

where we used the compact notation:

$$
\sum_\alpha \delta(f - f_\alpha) = e^{N\Sigma(f,T)}
\tag{2.82}
$$

and we evaluated the integral via the saddle-point method, provided that we are interested in the thermodynamic limit. Therefore we get:

$$
\left. \frac{\partial \Sigma(f, T)}{\partial f} \right|_{f^*(T)} = \frac{1}{T}
\tag{2.83}
$$

in the interval $[f_{min}, f_{th}]$, where Σ is the *complexity* or configurational entropy of the system. The main advantage of using mean-field p-spin models is the simple temperature dependence of the TAP free energy. Indeed, at zero temperature the states are simply identified by their energy, meaning that if we plot the complexity as a function of the energy density e, it is a concave function which goes to zero continuously at the ground state energy, e_{min}, and abruptly above a threshold value e_{th}. At finite temperatures the complexity is instead given by Eq. (2.82).

Three different regimes can be observed. (i) In the high-temperature phase, for $T > T_d$ the paramagnetic state dominates the free energy density for any value in the range $[f_{min}, f_{th}]$. This is nothing but the Gibbs state. (ii) For $T_K \leq T \leq T_d$, there is a value f^*, such that the quantity $f^* - T\Sigma(f^*)$ evaluated in f^* is equal to the paramagnetic solution f_{para}. In this second case, the paramagnetic state is given by an exponential number of metastable states of individual free energy density f^*. Despite the multiplicity of states, there is no phase transition because upon crossing T_d, the free energy preserves its analyticity without developing any singularity. (iii) A different situation takes place instead for $T < T_K$. The leading contributions are then provided by the lowest free energy states with $f^* = f_{min}$ and $\Sigma(f_{min}) = 0$. In this phase, $f = f_{min} - T\Sigma(f_{min}) = f_{min}$ because of the vanishing complexity $\Sigma(f_{min}, T_K)$, which is nothing but the configurational entropy at the Kauzmann point. In other words, the number of states is sub-exponential in the system size. To avoid a negative entropy paradox below T_K, the partition function is always assumed to be dominated by the same states with zero complexity.

While for $T > T_d$ the paramagnetic solution is always present, for $T < T_d$ it disappears and is replaced by a non-trivial combination of states. In the regime (T_K, T_d), the entropy is the sum of the complexity term and the vibrational contribution of a state labeled by f^*.

Given these premises, one might be interested in computing the complexity, which is in general a crucial quantity. However, sometimes one wants to grasp the physics behind the density of states $\Omega(f)$ without explicitly computing the TAP functional. In the following we will present two different methods, the first proposed by Monasson and the second by Franz and Parisi, also refereed to as *potential method*.

Let us focus first on the real replica method by Monasson [42], which consists in taking m copies of the same system and coupling them through a small parameter ϵ in the $\epsilon \to 0$ limit, after the thermodynamic limit. This attractive coupling is introduced to constrain the m copies to be in the same TAP state but, at the same time, they are uncorrelated within the state. Hence, the free energy is simply given by m times the contribution f_α in each state and the resulting partition function is:

$$Z_m \sim \sum_\alpha e^{-\beta N m f_\alpha} = \int_{f_{min}}^{f_{th}} df\, e^{N[\Sigma(f) - \beta m f]} \sim e^{N[\Sigma(f^*) - \beta m f^*]} . \qquad (2.84)$$

The saddle-point condition now verifies:

$$\frac{m}{T} = \left. \frac{\partial \Sigma(f, T)}{\partial f} \right|_{f^*(m,T)} . \qquad (2.85)$$

Thanks to the additional weight accounting for the m different copies, we can compute the full complexity function as follows:

$$\Phi(m, T) = -\frac{1}{\beta N} \log Z_m = \min_f [\beta m f - \Sigma(f)] = \beta m f^*(m, T) - \Sigma(f^*(m, T)),$$
(2.86)

from which we also have:

$$f^*(m, T) = \frac{\partial \Phi(m, T)}{\partial m},$$

$$\Sigma(f^*(m, T)) = m^2 \frac{\partial [m^{-1} \beta \Phi(m, T)]}{\partial m}.$$
(2.87)

Considering the parametric plot of $f^*(m, T)$ and $\Sigma(m, T)$, one can extrapolate the shape of $\Sigma(f)$ and consequently all the information about the TAP states.[19] The advantage of this method lies in the possibility to compute directly the complexity $\Sigma(f)$ at any temperature, once the free energy of m copies of the system and the analytical continuation to real values of m are known.

The second method developed by Franz and Parisi allows to define an effective potential, namely a large-deviation function $V(q)$ for the overlap parameter. The underlying hypothesis is that the system should be constrained to have a fixed overlap with a reference configuration $\mathbb{C}_{eq}$ extracted from the Boltzmann-Gibbs distribution at a temperature T [3, 21, 22]. In other words, $V(q)$ is the large deviation function of the probability to observe two configurations with overlap q. It is indeed related to the Parisi function $P(q)$ [43], i.e. $P(q) \simeq e^{-\beta N V(q)}$ at leading order. However, $P(q)$ cannot provide information on the dynamical transition, which the Franz-Parisi potential does.

The main goal of this framework is to analyze how much two configurations $\mathbb{C}$ and $\mathbb{C}_{eq}$ look similar, where the overlap $q(\mathbb{C}, \mathbb{C}_{eq})$ measures the degree of similarity. The partition function then reads:

$$Z(\mathbb{C}_{eq}, \epsilon, T) = \int d\mathbb{C} e^{-\beta H[\mathbb{C}] + \beta N \epsilon q(\mathbb{C}, \mathbb{C}_{eq})}$$
(2.88)

where ϵ is a coupling between the two configurations, $\mathbb{C}$ and $\mathbb{C}_{eq}$. The free energy $F(\mathbb{C}_{eq}, \epsilon, T)$ does not depend on the choice of the reference equilibrium configuration in the thermodynamic limit. Nevertheless, it constitutes a random object that needs be averaged over $\mathbb{C}_{eq}$:

$$F(\epsilon, T) = -\frac{T}{N} \int d\mathbb{C}_{eq} \frac{e^{-\beta H[\mathbb{C}_{eq}]}}{Z(T)} \log Z\left(\mathbb{C}_{eq}, \epsilon, T\right).$$
(2.89)

---

[19]Note that the value of m is either equal or less than one and this has different interesting implications.

The interesting issue consists in studying how possible correlations between the reference and the slave configurations affect the system, in the $\epsilon \to 0$ limit. The most natural way of interpreting this concept is to take the Legendre transform of the free energy:

$$V(q, T) = \max_{\epsilon} [F(\epsilon, T) + \epsilon q] \qquad (2.90)$$

where $q(\epsilon) = -\partial F(\epsilon, T)/\partial \epsilon$. More precisely, we obtain:

$$V(q, T) = -\frac{T}{N} \int d\mathbb{C}_{eq} \frac{e^{-\beta H[\mathbb{C}_{eq}]}}{Z(T)} \log Z \left(\mathbb{C}_{eq}, q, T\right) . \qquad (2.91)$$

The functional introduced here is analogous to that of ferromagnetic systems with long-range order. The difference is that in glasses there is no long-range order nor a simple order parameter that distinguishes all metastable states. In this specific case we actually focus on two real *replicas*, but the method can be safely generalized to three-replica potentials as proposed in [13, 23]. This approach is completely general, as $\mathbb{C}$ is susceptible to different interpretations. It can identify a spin configuration or equivalently a set of particle positions in glassy systems. In [22] the potential method was originally derived for the spherical p-spin model closer to the phenomenology of structural glasses. Such a model displays a first transition at T_d, at which the space of solutions splits into an exponential number of states with an extensive configurational entropy and another transition at T_K (a.k.a the *static transition temperature T_s*, already mentioned in Sect. 2.4). This scenario is in line with the RFOT theory.

Summarizing, there are two main classes defined on the basis of their replica symmetry breaking: (i) models with a continuous RSB, namely with a continuous Parisi order parameter $q(x)$, for which $T_d = T_K$; (ii) models with a single-step function (as the spherical p-spin model with $p > 2$), for which the dynamical transition occurs at a temperature higher than the static transition. In the second case the asymptotic state evolves far from its canonical average. In the $\epsilon \to 0$ limit the difference $V(q, T) - F(T)$ is qualitatively sketched in Fig. 2.6. Hence, at high temperatures a single phase exists and it coincides with the paramagnetic solution. The overlap parameter q is trivially zero since the reference configuration and the slave one belong to the same global minimum of the free-energy landscape with no correlation. Hence, the potential is convex with a unique minimum in $q = 0$. As temperature decreases many solutions appear, which correspond to metastable glassy states. This scenario is related to the appearance of a secondary local minimum q_2 at T_d at which the potential looks like a spinodal. The overlap value in this secondary minimum is the *self-overlap* of the TAP states. Further decreasing the temperature, at the static transition temperature the local minimum becomes the global one once the difference with the minimum located in $q = 0$ vanishes. Recalling the definitions given above, we thus have:

$$V(q_2, T) - V(0, T) = f^*(T) - F = f^*(T) - f^*(T) + T\Sigma(f^*(T)) = T\Sigma(f^*(T)) . \qquad (2.92)$$

Fig. 2.6 The Franz-Parisi potential as a function of the overlap q plotted for three different temperatures, $T > T_d$, $T = T_d$ and $T = T_K$. At T_d a dynamical transition occurs accompanied by the appearance of a secondary minimum. T_K is the aforementioned Kauzmann transition temperature associated with a vanishing configurational entropy

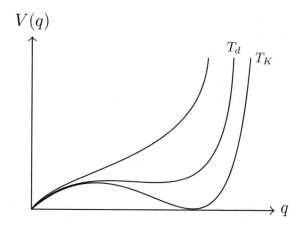

The equilibrium complexity $\Sigma(f^*(T))$ is exactly the logarithm of the number of metastable states that dominate the partition function in the *dynamical* phase and it goes to zero at the Kauzmann transition. For $T < T_K$ RSB effects can no longer be neglected, much complicating the study of the low-temperature regime. A related issue concerns the study of metastable TAP states in the intermediate temperature regime. In this case the potential will depend on two temperatures, T and T'. $V(q, T, T')$ is thus meant to be the minimal work to constrain the system, prepared at a temperature T, to be at a fixed overlap with a reference configuration of the same system at a different temperature T'. This approach is very instructive for the so-called *state following construction* developed later by the same authors [3, 39].

It is worth noting that the scenario described so far holds for mean-field models. In finite dimension, each state is expected to have a finite lifetime because of nucleation processes, as discussed above. Since the notion of true metastable state then loses its significance, in order to define a reasonable potential method one needs to turn to the Maxwell construction [16, 45].

References

1. Adam G, Gibbs JH (1965) On the temperature dependence of cooperative relaxation properties in glass forming liquids. J Chem Phys 43:139
2. Andreanov A, Biroli G, Bouchaud J-P (2009) Mode coupling as a Landau theory of the glass transition. EPL (Europhys Lett) 88(1):16001
3. Barrat A, Franz S, Parisi G (1997) Temperature evolution and bifurcations of metastable states in mean-field spin glasses, with connections with structural glasses. J Phys A Math General 30(16):5593
4. Bartenev GM (1966) Structure and mechanical properties of inorganic glasses. Moscow Publ. Lit. po Stroitelstvu
5. Bässler H (1987) Viscous flow in supercooled liquids analyzed in terms of transport theory for random media with energetic disorder. Phys Rev Lett 58:767

6. Bengtzelius U, Gotze W, Sjolander A (1984) Dynamics of supercooled liquids and the glass transition. J Phys C Solid State Phys 17(33):5915
7. Berthier L, Biroli G (2011) Theoretical perspective on the glass transition and amorphous materials. Rev Mod Phys 83:587
8. Biroli G, Bouchaud JP (2007) Critical fluctuations and breakdown of the Stokes-Einstein relation in the mode-coupling theory of glasses. J Phys Cond Matt 19:205101
9. Biroli G, Bouchaud JP (2012) The random first-order transition theory of glasses: a critical assessment. In: Wolynes PG, Lubchenko V (eds) Structural glasses and supercooled liquids: theory, experiment and applications. Wiley
10. Bouchaud J-P, Biroli G (2004) On the Adam-Gibbs-Kirkpatrick-Thirumalai-Wolynes scenario for the viscosity increase in glasses. J Chem Phys 121:7347
11. Castellani T, Cavagna A (2005) Spin glass theory for pedestrians. J Stat Mech: Theory Exp 2005(05):P05012
12. Cavagna A (2009) Supercooled liquids for pedestrians. Phys Rep 476(4–6), 51
13. Cavagna A, Giardina I, Parisi G (1997) An investigation of the hidden structure of states in a mean-field spin-glass model. J Phys A Math General 30(20):7021
14. Cavagna A, Giardina I, Parisi G (1998) Stationary points of the Thouless-Anderson-Palmer free energy. Phys Rev B 57(18):11251
15. Chandler D, Garrahan JP (2010) Dynamics on the way to forming glass: bubbles in space-time. Ann Rev Phys Chem 61:191–217
16. Clerk-Maxwell J (1875) On the dynamical evidence of the molecular constitution of bodies
17. Crisanti A (2008) Long time limit of equilibrium glassy dynamics and replica calculation. Nuclear Phys B 796(3):425
18. Fischer KH, Hertz JA (1991) Spin glasses (Cambridge Studies in Magnetism)
19. Flory P-J (1956) Statistical thermodynamics of semi-flexible chain molecules. Proc R Soc Lond A 234(1196):60
20. Franz S, Montanari A (2007) Analytic determination of dynamical and mosaic length scales in a Kac glass model. J Phys A Math Theor 40(11):F251
21. Franz S, Parisi G (1997) Phase diagram of coupled glassy systems: a mean-field study. Phys Rev Lett 79(13):2486
22. Franz S, Parisi G (1995) Recipes for metastable states in spin glasses. J Phys I 5(11):1401
23. Franz S, Parisi G, Virasoro MA (1993) Free-energy cost for ultrametricity violations in spin glasses. EPL (Europhys Lett) 22(6):405
24. Fredrickson GH, Andersen HC (1984) Kinetic ising model of the glass transition. Phys Rev Lett 53(13):1244
25. Gallavotti G (2013) Statistical mechanics: a short treatise. Springer Science & Business Media
26. Garrahan JP, Chandler D (2003) Coarse-grained microscopic model of glass formers. Proc Natl Acad Sci 100(17):9710
27. Gibbs JH, DiMarzio EA (1958) Nature of the glass transition and the glassy state. J Chem Phys 28(3):373
28. Götze W (1985) Properties of the glass instability treated within a mode coupling theory. Z Phys B Condens Matter 60(2–4):195
29. Hansen IR, McDonald J-P (1990) Theory of simple liquids. Elsevier
30. Itzykson C, Drouffe J-M (1991) Statistical field theory: volume 2, strong coupling, Monte Carlo methods, conformal field theory and random systems, vol. 2. Cambridge University Press
31. Jäckle J, Eisinger S (1991) A hierarchically constrained kinetic ising model. Z Phys B Condens Matter 84(1):115
32. Johari GP (2000) A resolution for the enigma of a liquid's configurational entropy-molecular kinetics relation. J Chem Phys 112(20):8958
33. Kirkpatrick TR, Thirumalai D (1988) Mean-field soft-spin Potts glass model: Statics and dynamics. Phys Rev B 37(10):5342
34. Kirkpatrick TR, Wolynes PG (1987) Stable and metastable states in mean-field Potts and structural glasses. Phys Rev B 36(16):8552

35. Kob W (2002) Supercooled liquids, the glass transition, and computer simulations. arXiv preprint arXiv:cond-mat/0212344
36. Kob W, Andersen HC (1993) Kinetic lattice-gas model of cage effects in high density liquids and a test of mode-coupling theory of the ideal-glass transition. Phys Rev E 48(6):4364
37. Kob W, Andersen HC (1995) Testing mode-coupling theory for a supercooled binary Lennard-Jones mixture I: the van Hove correlation function. Phys Rev E 51(5):4626
38. Kohlrausch R (1847) Ann Phys (Leipzig) 12:393
39. Krzakala F, Zdeborová L (2010) Following Gibbs states adiabatically-The energy landscape of mean-field glassy systems. EPL (Europhys Lett) 90(6):66002
40. Laughlin WT, Uhlmann DR (1972) Viscous flow in simple organic liquids. J Phys Chem 76(16):2317
41. Leuzzi L, Nieuwenhuizen TM (2007) Thermodynamics of the glassy state. Taylor & Francis
42. Monasson R (1995) Structural glass transition and the entropy of the metastable states. Phys Rev Lett 75(15):2847
43. Parisi G, Mézard M, Virasoro MA (1987) Spin glass theory and beyond. World Scientific, Singapore
44. Parisi G, Zamponi F (2010) Mean-field theory of hard sphere glasses and jamming. Rev Mod Phys 82(1):789
45. Reichl LE (1999) A modern course in statistical physics
46. Reichman DR, Charbonneau P (2005) Mode-coupling theory. J Stat Mech: Theory Exp 2005(05):P05013
47. Ritland HN (1954) Density phenomena in the transformation range of a borosilicate crown glass. J Am Ceramic Soc 37(8):370
48. Ritort F, Sollich P (2003) Glassy dynamics of kinetically constrained models. Adv Phys 52(4):219
49. Royall CP, Williams SR (2015) The role of local structure in dynamical arrest. Phys Rep 560:1
50. Schwarz JM, Liu AJ, Chayes LQ (2006) The onset of jamming as the sudden emergence of an infinite k-core cluster. EPL (Europhys Lett) 73(4):560
51. Sompolinsky H, Zippelius A (1982) Relaxational dynamics of the Edwards-Anderson model and the mean-field theory of spin-glasses. Phys Rev B 25(11):6860
52. Szamel G (2003) Colloidal glass transition: beyond mode-coupling theory. Phys Rev Lett 90(22):228301
53. Szamel G (2010) Dynamic glass transition: bridging the gap between mode-coupling theory and the replica approach. EPL (Europhys Lett) 91:56004
54. Thirumalai D, Kirkpatrick TR (1988) Mean-field Potts glass model: initial-condition effects on dynamics and properties of metastable states. Phys Rev B 38(7):4881
55. Thouless DJ, Anderson PW, Palmer RG (1977) Solution of 'solvable model of a spin glass'. Philos Mag 35(3):593
56. Toninelli C, Biroli G, Fisher DS (2006) Jamming percolation and glass transitions in lattice models. Phys Rev Lett 96(3):035702
57. Williams G, Watts DC (1970) Non-symmetrical dielectric relaxation behaviour arising from a simple empirical decay function. Trans Faraday Soc 66:80
58. Zamponi F (2010) Mean field theory of spin glasses. arXiv preprint arXiv:1008.4844
59. Zwanzig R (2001) Nonequilibrium statistical mechanics. Oxford University Press, Oxford

Chapter 3
The Jamming Transition

In Chap. 2 we presented a rather broad overview of renowned theories about the glass transition. In the last decades, the study of glasses at low temperature has also attracted significant interest, both from a theoretical and an experimental point of view [29, 39, 40, 42]. This is motivated by the fact that a different *solidity* transition, the *jamming transition*, can be observed in hard spheres and in many other out-of-equilibrium systems, such as foams, emulsions, granular matter.

The jamming phenomenon defines a purely geometrical problem where the thermal energy does not play any role in determining or facilitating the transition. It has been studied for years but only recently new advancements have been achieved by analyzing systems of granular media. Liu and Nagel first proposed a compelling unifying phase diagram [40] in an attempt to group molecular glasses, colloids, foams, emulsions, hard spheres—in one word glassy systems together with granular ones— and connecting various rigidity transitions. For instance, granular media flow when they are shaken or poured, but jam below a certain shaking intensity. Analogously, foams and emulsions have a liquid-like behavior for a large shear stress, but jam when the shear stress is lowered below the yield stress [20]. However, while colloids and molecular liquids are both thermal systems, granular media, dense emulsions, foams and hard spheres, are athermal. For the last category the interaction energies are orders of magnitude larger than the thermal scale $k_B T$. Of course, the determination of a unique phase diagram might be helpful to investigate different classes of materials and the different role played by thermal fluctuations. However, one of the main difficulties in studying jammed systems is that, while thermodynamics usually provides a unique phase (or a finite number of phases) at equilibrium, defining a single phase in out-of-equilibrium systems looks unfeasible.

© Springer Nature Switzerland AG 2019
A. Altieri, *Jamming and Glass Transitions*, Springer Theses,
https://doi.org/10.1007/978-3-030-23600-7_3

3.1 Theoretical and Numerical Protocols in Jammed Systems

The basic jamming scenario takes place in models of frictionless spheres interacting via finite-range repulsive forces at zero temperature. This might be considered as the building block, the Ising model for jamming. The locus of points J, which define the jamming transition, depicts a new physics, which in some aspects resembles an ordinary second-order phase transition for which the scaling theory of critical phenomena [32] applies. There are, however, other features that explicitly deviate from a second-order transition behavior.

3.1.1 Edwards Conjecture: Equiprobability of Jammed Configurations

First, it is important to identify the macroscopic variables of interest for the jamming transition (Fig. 3.1). The energy, for instance, can be conserved or not, but it does not play any role for hard-sphere configurations. Alternative control parameters might be either the packing fraction or the system volume, which will depend on the grain configurations. This was the main idea on which Edwards built his insight, proposing

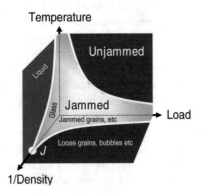

Fig. 3.1 Figure reprinted from [39], showing that the J-point is at the boundary of the jammed region at zero temperature and zero applied shear stress. This point can be clearly identified in frictionless systems with repulsive, finite-range interactions. Materials that undergo a jamming transition at low temperatures and high density, can unjam and yield if the control parameters are appropriately changed. This is just an indicative plot, as the shape (typically not so sharp) of the jamming line can be different for different systems. T and 1/Density are usually employed as reference axes. The glass transition picture lives in the vertical plane: it is identified by the curve which separates the jammed phase from the unjammed one in the 1/Density-T plane. Foams and emulsions instead usually live on the horizontal 1/Density-Σ plane reaching the J-point for zero shear

Fig. 3.2 Five grains of generic shape are pictured in a jammed configuration. *a* represents the contact, while *i* labels the particle. The grey particle in the center is stable if the two last conditions in Eq. (3.1) are verified. Figure taken from [3]

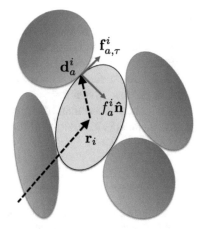

to parametrize the jammed state throughout a volume function $\mathcal{W}\left(\{\vec{r}_i, \vec{\hat{t}}_i\}\right)$, which depends on the particle positions $\vec{r}_i$ and their orientations $\hat{t}_i$ [21, 22]. According to this description, only jammed states are considered, being a jammed state built on the excluded volume principle and on force and torque balance constraints. For an assembly of N monodisperse hard spheres of equal radius r these two constraints translate into [3]:

$$\begin{cases} \left|\vec{r}_i - \vec{r}_j\right| \geq 2r \\ \sum_{a \in \partial i} \vec{f}_a^i = 0 \quad i = 1, ..., N \\ \sum_{a \in \partial i} \vec{d}_a^i \times \vec{f}_a^i = 0 \,. \end{cases} \tag{3.1}$$

The notation ∂i stands for the set of contacts of particle i, while $\vec{d}_a^i$ is the vector drawn from the center of particle i to the contact point a (Fig. 3.2). For this contact there is a force vector $\vec{f}_a^i$. The last two conditions remain valid for frictional particles with the only *caveat* to decompose the force $\vec{f}_a^i$ into normal and tangential components. If all forces and torques on the particles are perfectly balanced then mechanical stability is exactly achieved. For purely repulsive particles, one should also verify: $\vec{d}_a^i \cdot \vec{f}_a^i < 0$.

Since in this framework the volume seems to be the most relevant thermodynamic descriptor, one can define a measure of all possible microstates as:

$$S(V) = \rho \log \Omega(V) \,, \quad \Omega(V) = \prod_{i=1}^{N} \int d\vec{r}_i \oint d\vec{\hat{t}}_i \delta(V - \mathcal{W}\left(\{\vec{r}_i, \vec{\hat{t}}_i\}\right)) \Theta_J \,, \quad (3.2)$$

where Θ_J encompass all the conditions reported in Eq. (3.1). The definition of Edwards entropy $S(V)$ used in condensed matter corresponds to the analogous concept of configurational entropy introduced in Chap. 2 and used in spin glasses and structural glasses. The main message emerging from Eq. (3.2) and summarized in

one sentence is that, according to Edwards conjecture, mechanically stable configurations are equally likely. Several numerical studies have been carried out in the
last years to test this conjecture, confirming it and offering at the same time new
interesting perspectives [43].

Let us browse rapidly some of the main techniques used to investigate the properties of jammed systems.

In a numerical simulation one can generally use the following definition of the
potential, suitable to describe different kinds of interactions:

$$V(r_{ij}) = \begin{cases} \frac{\epsilon}{\alpha} \left(1 - r_{ij}/\sigma_{ij}\right)^{\alpha} & \text{for} \quad r_{ij} < \sigma_{ij} \\ 0 & \text{for} \quad r_{ij} \geq \sigma_{ij} \end{cases} \tag{3.3}$$

where ϵ is an energy scale, r_{ij} the mutual distance between the centers of particles i
and j and σ the sum of their radii. Starting from this generic form, one might study
three different cases: (i) $\alpha = 2$, which identifies harmonic springs; (ii) $\alpha = 3/2$ for
non-linear springs; (iii) $\alpha = 5/2$ for Hertzian interactions with a softer effect than
(i). All these cases refer to repulsive and finite-range interactions. This choice is
typically motivated by granular systems for which particles do not interact unless
they overlap. Why is using packing of soft frictionless spheres so convenient? First,
their jamming line is well-defined, with a clear transition threshold from a zero to a
finite pressure. Second, intriguing properties emerge as a consequence of the direct
correspondence between the jamming point and a marginal stability condition that we
will explain in more detail in the following. Last, but not least, physical observables
exhibit non-trivial power-law scalings as a function of the distance from the jamming
threshold.

Simulations on potentials like (3.3) usually involve two-dimensional systems
with bidisperse particles (a usual trick to avoid crystallization), polydisperse or
monodisperse three-dimensional systems and molecular dynamics techniques. However, understanding how configurations at $T = 0$ can be correctly generated is a central issue. Typically one starts considering N particles whose positions are chosen
completely at random—in order to reproduce the $T \to \infty$ limit and to sample the
space uniformly—in a given box in $2d$ or $3d$ with periodic boundary conditions. Then
using conjugate-gradient techniques [51], one lets the system evolve to the nearest
minimum of the potential energy landscape. The minimization procedure ends when
prescribed criteria are verified, i.e. when the total potential energy per particle is less
than a certain fixed threshold and its error in next iterations is less than a fixed value,
say 10^{-15} [47]. A genuine problem that arises in this scheme is that each initial state
in the $T \to \infty$ limit can generate a different value of the critical packing fraction
ϕ_c, the jamming onset. For this reason, it is more appropriate to measure observables as a function of the excess density $\phi - \phi_c$, which instead provides more robust
outcomes. The distribution of jamming thresholds $P(\phi_c)$ in infinite-size systems is
nevertheless shown to be independent of the choice of α in the potential expression.
Another remarkable property concerns the estimated value of the threshold, which
turns out to be really close to the corresponding *random close packing* RCP, that

is for instance in a monodisperse three-dimensional system $\phi_{RCP} \approx 0.64$ [4, 59]. Obviously, the random close packing density in ordered systems is higher than the corresponding value in disordered ones: order packings are known to be denser than the analogous irregular ones [60]. However, the notion *random close packing* might be a bit misleading as it involves two very different concepts at the same time, i.e. randomness and dense packing. This is why Torquato et al. [58] suggested to use the alternative definition of *maximally random jammed state*.

There are different methods to determine numerically the random close packing density, which nevertheless should be carefully considered in the case of hard spheres, corresponding to $\epsilon \to \infty$. The main difficulties stem from the definition of a suitable algorithm for dealing with hard-sphere configurations. One of the most accepted methods is based on *density ramps* as suggested by Lubachevsky and Stillinger [41].[1] The usual way to proceed consists in starting from a random configuration of spheres in a given volume with fixed boundary conditions. The radius of the particles is expanded uniformly at a rate λ during a molecular dynamics run. Thanks to this inflating process particles can collide and finally jam, for a diverging collision rate. On the one hand, if λ goes to zero, the system is indefinitely in equilibrium and tends to crystallize. On the other hand, for λ positive but sufficiently small, the system can reach a jammed state. The procedure is repeated several times in order to have a well-defined density of states. Most of these procedures lead to packing fractions close to 0.64 in three dimensions and to 0.84 in two dimensions, if crystallization is avoided.

Generally, the definition of amorphous states requires solving complicated non-equilibrium dynamics, as in the Lubachevsky-Stillinger algorithm. An alternative way would be to identify the different classes to which amorphous packings belong, defined as the infinite pressure limit of glassy states. This scenario can be understood using exclusively equilibrium statistical mechanics methods.

Let us give some intuitions with a particular emphasis on systems formed by frictionless hard spheres. In Fig. 3.3 we show the equilibrium phase diagram, from which to study the metastable liquid branch we need to assume that the equation of state of the liquid can be continued above ϕ_f, the freezing density [2, 33]. However, it is a quite strong assumption: it can be deflated in the presence of a kinetic spinodal [8], which prevents the system from reaching amorphous states if the compression rate is not sufficiently high. Several inconsistencies can be highlighted, as the prediction of unphysical values for the density where the pressure should diverge, or even an ambiguity in defining the continuation of the liquid equation of state. However, improving the analysis by means of a refined version of the so-called Percus-Yevick equations [49], a phase transition at high density is reasonably expected [53].

To overcome these problems, one might propose an alternative viewpoint based on the existence of a thermodynamic phase transition in the metastable branch of Fig. 3.3, at ϕ_K. This transition is also predicted in mean-field models [5, 16, 52] and discussed

---

[1]According to this protocol, the energy landscape changes step by step with the density.

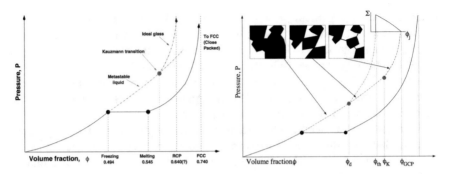

Fig. 3.3 Phase diagram of a system of hard spheres in three dimensions. The full black line is the equilibrium line that, above a freezing density value, displays a transition towards the crystalline phase. The most stable structure around which particle vibrate is given by the FCC lattice. The dashed line represents the continuation of the liquid branch, assuming to be able to avoid crystallization and to enter the metastable branch. The point where, upon compressing the system, the pressure diverge is the random close packing density (RCP). On the right, an analogous mean-field phase diagram is shown where on the horizontal axis the most relevant densities for the metastable states are reported. At ϕ_d a clustering transition occurs, characterized by the appearance of an exponential number of states. At ϕ_K, called Kauzmann packing, a thermodynamic glass transition takes place. Compressing each state sufficiently fast, one can produce jammed packings, from ϕ_d up to ϕ_{th} and from ϕ_K up to ϕ_{GCP}. The inset shows the configurational entropy of these metastable states as a function of the jamming density, while the three boxes on the left show the allowed configurations (in black) compared to the forbidden ones (in white). The figures are taken from [48]

in Chap. 2 concerning the MCT section.[2] The signature of the thermodynamic glass transition is a jump in the compressibility of the system. In fact, in the dense liquid two time scales can be observed: one due to particle vibrations inside cages, and a second slower, due to cooperative rearrangements. Changing the density by an amount $\Delta\phi$, a corresponding increase in the pressure ΔP_0 should occur, accompanied by the reduction of the average size of the cages. The pressure then relaxes to $\Delta P_f < \Delta P_0$. On larger time scales, due to structural rearrangements, the pressure further relaxes to a value $\Delta P_\infty < \Delta P_f$. Generally, one assumes that at the thermodynamic glass transition the structural relaxation time is infinite and then the process is essentially frozen. This phenomenon implies that the increase in pressure is larger in the glassy phase than in the metastable liquid phase that yields a smaller compressibility $K = \phi^{-1}\left(\frac{\Delta\phi}{\Delta P}\right)$. This causes a jump in the compressibility.

The scenario described above, conjectured to be relevant in finite dimension, is generally complicated in mean-field systems or infinite dimensions, since before the thermodynamic glass transition a *clustering transition* is expected. It corresponds to the appearance of an exponential number of metastable states. Above this density, the emerging metastable states correspond to locally stable configurations around which

---

[2]More precisely, MCT predicts a divergence of the relaxation time at $\phi_{MCT} < \phi_K$. At the mean-field level, this transition coincides with ϕ_d where an ergodicity breaking against the single-state liquid phase occurs.

the system vibrates. Hence, while at ϕ_d the first metastable glassy states arise, at ϕ_K an ideal thermodynamic glass transition, with an associated jump in the compressibility, takes place. Between ϕ_d and ϕ_K another category of states can be detected, which can be followed by compressing the system fast enough to avoid structural relaxation. Each of the states can be associated with a final jamming density, ϕ_j, with a particular emphasis on ϕ_{GCP}, i.e. the glass close packing density, and ϕ_{th}, i.e. the threshold density corresponding to the less dense glassy states.[3] The number of states, which correspond to a jamming density ϕ_j, increase exponentially with system size, namely $\mathcal{N}(\phi_j) = \exp\left[N\Sigma(\phi_j)\right]$, where again $\Sigma(\phi_j)$ represents the system complexity.

It is worth highlighting that ϕ_{GCP} does not depend on the specific protocol. The interesting feature, which opens the door to a sort of universality principle, is that some properties characterize all jammed amorphous packings per se, irrespective of numerical setup.

3.2 The Marginal Glass Phase

The transition previously discussed has an interesting interpretation in dynamical terms. Indeed, while in the liquid phase, all regions of phase space can be dynamically reached, in a normal glass, the space of solutions splits into an exponential number of metastable states. This breaking point corresponds to the dynamical critical density already mentioned in terms of dynamical temperature T_d. The normal glass defines an *entropically rigid state*, in the sense that if one starts by considering a given set of configurations, only motion within that set is allowed (see the second phase in Fig. 3.4). The mean-square displacement of the particles in this regime has a long-time limit that is proportional to the amplitude of the vibrations inside a cage. Then, we can recognize a third phase, where the system is expected to be *marginally stable*. In this region, those we previously described as minima are metabasins, or in other words an ensemble of metastable states, each of which further splits according to an ultrametric structure. This scenario is exactly shown in Fig. 3.4. The last phase is characterized by a non-stationary dynamics at long times with the emergence of multiple plateaus and highly heterogenous spatial vibrations. Let us try to clarify better the main differences between these three phases. In the liquid phase we can straightforwardly introduce a potential method which is well-defined everywhere without any local minimum. By following the liquid line for increasing densities, above a certain threshold (the *dynamical density*), the system becomes frozen: the diffusion constant drops to zero and many glassy states appear. The pressure of these new states nevertheless analytically continues that of the equilibrium liquid analyzed at that density. Therefore, the thermodynamic is well-defined and perfectly smooth, while the dynamics is stuck.

---

[3]Note that ϕ_{th} is not exactly $\phi_j(\phi_d)$. Both are unique functions of the initial equilibrium density, but obtained in different ways. A similar argument holds for ϕ_{GCP}. An extensive explanation might be found in [11].

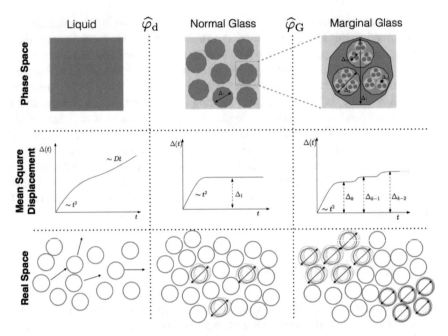

Fig. 3.4 The picture shows the space of solutions, the mean-square displacement and the arrangement of particles in real space from top to bottom. In the liquid phase (left) the dynamics is diffusive and the spheres are completely free in space. In the normal glass (center), the dynamics is no longer diffusive and the mean-square displacement is bounded from above having a finite long-time limit. In the marginal glass (right), the dynamics shows different plateaus at different heights. In real space particles vibrate in a special manner, giving rise to highly heterogenous vibrations. Figure taken from [11]

What happens at very high densities or pressures? This question opens new interesting scenarios, as each glass state at a certain point will undergo a *Gardner phase transition*, for which any simple interpretation proposed before ceases to hold. This transition was proposed first in spin-glass models by Gardner [26] and by Gross et al. [28] in the same year. This aspect corroborates again the strict connection between structural and spin glasses, in their formalism and in their underlying physical understanding.

In correspondence of the Gardner density each glass state displays a complex internal structure, modifying its nature from a basin to a metabasin.

Going deeply into the low-energy phase, beyond the Gardner point, a marginal glass phase takes place. We will see in the following what it means in terms of vibrational modes. When further compressing the system, the endpoint of the compression protocol coincides with the *jamming onset*, signaled by a diverging value of the pressure. The spheres are all in contact and the densest packing point is finally reached. The jamming transition is then a non-trivial phenomenon occurring in the large-pressure limit inside the marginal glass. Even though a strong protocol dependence affects the final density ϕ_j, important properties concerning the distributions of weak forces and small gaps between the particles appear quite robust and universal. Critical exponents for these distributions obtained in numerical simulations for dimensions $d = 2$ up to $d = 12$ turn out to be in remarkable agreement with the analytical predictions in $d \to \infty$ [12, 13]. Surprisingly, this universal trend seems to be a direct consequence of the hierarchical structure of sub-basins.

3.3 Anomalous Properties of Jammed Systems

In the following Section we aim at briefly describing some of the main properties that characterize the jamming transition, with a particular emphasis on the concept of *isostaticity* and unexpected anomalous behaviors in the distribution of soft modes.

3.3.1 Isostaticity

First, we focus on the concept of *isostaticity*. Since jamming takes place at $T = 0$, the system is always in mechanical equilibrium. Thus, forces are perfectly balanced on all particles. Given N soft spheres in d dimensions, the number of equations to satisfy is Nd. Moreover, as Maxwell claimed [45], the rigidity condition requires that the number of inter-particle forces $\frac{Nz}{2}$ (where z stands for the average connectivity per particle and each contact is shared at most by two particles) must be at least equal to the number of force balance equations, i.e. Nd. This condition leads to $z \geq 2d$.

In the jamming limit the distance between particles in contact must be exactly equal to the sum of their radii, according to the *just-touching* condition. This implies $\frac{Nz}{2}$ constraints on the positional degrees of freedom that are Nd. A generic solution is possible if $z \leq 2d$. Gathering all the information, the only reasonable solution exactly at jamming is[4] $z_c = 2d$. Note that the equality condition between the total number of constraints and the number of degrees of freedom is also due to the absence of *floppy modes*, responsible for deformations in the systems with zero energy cost in lowest order. They would be present if the system was not sufficiently connected, as Maxwell suggested. Indeed, when the system has a low average coordination number, these soft modes emerge as a signature of non-rigidity.

Another tricky aspect that typically arises in numerical simulations is the presence of *rattlers*, particles that do not have any contact. Treating correctly this type of particles and eventually removing them is important, albeit they do not contribute to jamming. A technical problem related to rattlers is that they are much faster than other particles and have their own peculiar dynamical properties. Thus, to properly study the system dynamics, one can either define a modified version of the mean-

---

[4]Note that this condition holds in the case of frictionless, spherical particles with purely repulsive interactions. In fact, frictional spheres are almost always hyperstatic at the jamming transition, with $z > z_c$, as not only normal forces between the particles are taken into account.

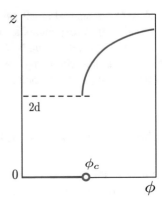

Fig. 3.5 Qualitative plot of the coordination number as a function of density. The Figure, reprinted from [30], displays a jump at the critical density ϕ_c

square displacement, which does not account for particles that move more than the average value [31],[5] or remove rattlers from the very beginning and compute the mean squared displacement on $N' < N$ particles that are not rattlers at all. Especially in the unjammed regime, for density $<\phi_c$, rattlers may spoil measurements, so that appropriate techniques are then fundamental.

3.3.2 Coordination Number

As mentioned above, the jamming transition displays both first order and second order transition behaviors. Similarly to a first-order transition, a discontinuous jump in the coordination number z occurs at jamming from zero to $2d$, as shown in Fig. 3.5. Due to the mixed nature of the jamming transition, a power-law scaling is also observed in the excess contact number $z - z_c$:

$$z - z_c = z_0 \, (\phi - \phi_c)^\zeta \tag{3.4}$$

where $\zeta = 0.50 \pm 0.03$ from numerical fits and the precise value of z_0 depends on the dimension of the system. The exponent ζ does not depend on the specific expression of the potential, being related only to the geometry of the packing. In this case, to reproduce the correct scaling between z and the excess density, one needs to exclude rattlers in the contact count but re-include them for the packing fraction calculation [47]. Note that above the jamming threshold the condition $z_c = 2d$ is no longer valid and additional equations in terms of force balance are needed.

Several other quantities follow a power-law scaling as a function of the distance from jamming, such as the pressure, the static shear modulus G_∞ and even the peak

---

[5]To study vibrational motion the authors suggest to use a different definition: $\tilde{\Delta}^2(t) = \frac{1}{N}\sum_{i=1}^N \frac{3}{\langle|\Delta\vec{r}_i(t)|^{-2}\rangle}$ in such a way to have exactly the standard definition of mean-square displacement if the single-particle distribution is Gaussian.

position in the distribution of the jamming threshold. The key point is that the exponents that govern the power-law scalings are all independent of dimensionality and polydispersity. The independence on dimension down to $d = 2$ seems to corroborate the hypothesis that the upper critical dimension for the jamming transition is exactly equal to two [40, 62]. More tests should, however, be done in this direction to have a more robust evidence of it.

3.3.3 Density of States

When the contact number z decreases below a certain value, the system loses its rigidity. In terms of vibrational properties this corresponds to the emergence of collective particle motions, the aforementioned floppy modes, i.e. modes that do not require any elastic energy cost. Compared to ordinary solids with long-range crystalline order, the spectrum of low-energy excitations in jammed materials exhibits several anomalies [63, 64, 66]. These anomalous properties cannot be ascribed to the elasticity theory but rather to a completely new physical scenario. Therefore we need to analyze the response of the system against weak quasi-static perturbations and to compute the normal modes and their eigenfrequencies. The starting point is the definition of the dynamical matrix of the jammed packing, whose diagonalization provides the squared frequencies $\lambda = \omega^2$ and the polarization vectors in each mode. Taking into account only linear deformations, we introduce the relative displacement $\vec{u}_{ij}$ of two close particles i and j. The energy difference is due to:

$$\Delta E = \frac{1}{2} \sum_{i,j} k_{ij} \left(u_{\parallel,ij}^2 - \frac{f_{ij}}{k_{ij} r_{ij}} u_{\perp,ij}^2 \right) \tag{3.5}$$

where k_{ij} is the stiffness coefficient and f_{ij} the force between i and j. The displacement u_{ij} has been decomposed in its parallel and perpendicular components to $\vec{r}_{ij}$, where the latter represents the mutual distance between the centers of the same particles. Rewriting the expression for the energy variation in terms of independent vectors, we get:

$$\Delta E = \frac{1}{2} \mathcal{M}_{ij,nm} u_{i,n} u_{j,m} , \tag{3.6}$$

in which the definition of the dynamical matrix appears, i.e. a square $dN \times dN$ matrix. The convention of summing over repeated indices is implicitly used.

Low-frequency modes in ordinary solids are long-wavelength plane waves, whose density of states (DOS) scales as $D(\omega) \sim \omega^{d-1}$, ω being the frequency of the modes and d the dimension of the system in exam. A striking feature characterizing amorphous solids is the violation of the expected Debye law close to jamming. This feature is generally refereed to as *boson peak* because the ratio of the observed DOS and the expected Debye law shows a peak at low frequencies. Since its discovery, numerous studies have been carried out, although leading to very different

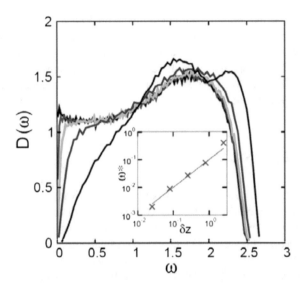

Fig. 3.6 Density of states $D(\omega)$ as a function of the frequency ω reproduced from simulation data in [47], considering order one thousand of soft spheres in a periodic cubic box subject to a repulsive harmonic potential. The distance from jamming is progressively reduced from $\Delta\phi = 0.1$ (black) to 10^{-4} (red). The inset shows the cut-off frequency ω^*, corresponding to the point where $D(\omega)$ is half of its plateau value, as a linear function of the excess contact number

and sometimes divergent explanations. According to one of these, the boson peak would only be a generalized version of Van Hove singularities occurring in crystals [57]. Far from jamming, amorphous solids and crystal look alike, as for instance in three-dimensional systems the DOS scales as ω^2 in both cases. What makes the real difference is the nature of the anomalous modes emerging near jamming, which are a characteristic signature of weakly-connected amorphous solids (Fig. 3.6). As briefly mentioned above, when any contact is removed in the system, a floppy mode naturally arises with an associated zero frequency. It is possible to relate the emergence of the plateau in $D(\omega)$, and hence the excess of low-frequency modes, to a specific set of floppy modes. If one considers an ordinary solid, it has three soft modes that are associated with translational motion along the three coordinate axes. If the system is constrained in a rigid box, translation can no longer be considered a soft mode. To recover the fundamental low-frequency modes one can either apply a sinusoidal or a similarly smooth distortion of the original soft modes. The same procedure can be carried out in an isostatic packing [63].

However, in an amorphous solid close to jamming, the number of floppy modes is greater than three, typically of order L^{d-1} for a regular cubic box of linear size L. The density of modes per unit volume is $D(\omega_L) \sim (L^{d-1}/L^{-1})L^{-d} \sim L^0$ normalized to the leading frequency generated by a distortion, i.e. $\omega_L \sim L^{-1}$. This means that the density of these anomalous modes tends to a non-vanishing constant value even as $\omega_L \to 0$. The crossover between the two regimes, the one for which the Debye predictions continue to apply, and the other characterized by an atypical behavior, is signaled by the characteristic value ω^*.[6] This crossover frequency, generally belonging to the terahertz range, scales as $\omega^* \sim \Delta z = z - z_c$ and it is strictly related to diverging correlation lengths [56, 62, 64] of great interest for the jamming transition.

Hence, the non-vanishing constant value that the density of states reaches, instead of dropping to zero as expected for sound modes, is properly due to an excess of low-frequency modes, which can be also detected either in numerical simulations or in Raman and neutron scattering experiments. First insights on this phenomenon date back to [6, 7, 27, 44]. A related issue concerns the properties of these normal modes, which are highly heterogeneous and resonant near ω^* and become quasi-localized upon decreasing the frequency [65]. Above the crossover frequency, a broad range of wavevectors contributes to making the vibrational modes more and more extended. To characterize their spread and determine whether they are localized or not, one can use several order parameters, such as the inverse participation ratio (IPR) $\mathcal{I}$, the level repulsion[7] and localization length [55, 67]. The IPR is defined as:

$$\mathcal{I} = \frac{1}{N} \frac{\left(\sum_i |u_i|^2\right)^2}{\sum_i |u_i|^4} , \tag{3.7}$$

where u_i is the polarization vector of particle i. Depending on its order of magnitude, one can observe extended modes, if $\mathcal{I} \sim O(1)$, or localized modes, if $\mathcal{I} \sim O(1/N)$, meaning that there is only one region of the system that has one major contribution.

From a broader perspective, it is however unquestionable that the anomalous distribution of modes has non-trivial implications even for quantum problems, concerning, in particular, the behavior of the thermal conductivity and the specific heat. In particular, the specific heat can be expressed as a sum of a cubic term and of a linear one:

$$c_v \approx A_{\text{Debye}} T^3 + BT . \tag{3.8}$$

The first term is due to the ordinary Debye behavior of long-wavelength modes, while the latter arises because of new modes, which can be typically described within two-level tunneling models [1, 50]. It is an immediate conclusion that even without assuming tunneling excitations, a flat distribution in the density of states should naturally lead to a linear contribution in the specific heat.

A full theory that further explores and explains these low-energy excitations is in progress. To mention some recent advancements in this direction, the replica method

---

[6]Falling down in the terahertz range a priori the boson peak should not have relevance for the glass transition. However, it has been shown that materials with a pronounced boson peak belong to strong glasses, while those with a weak increase are mostly fragile. Possible connections between this low-frequency effect and instabilities characterizing the glass transition have also been highlighted in colloids [15].

[7]Typically, low-frequency modes are mostly extended and exhibit level repulsion, i.e. their level spacing statistics follows a *Wigner surmise* in random matrix theory, while high-frequency modes are localized and exhibit Poissonian level statistics.

has been employed for instance to study static properties of the glassy phases, [9, 24, 35, 36], allowing an analytical derivation of characteristic scalings and critical exponents in agreement with the values observed in numerical simulations [12, 14]. The analysis has also been extended to the dynamics in order to derive Mode-Coupling-like equations for the correlation functions [34].

3.3.4 Diverging Correlation Lengths

The vanishing trend of ω^* upon approaching the jamming threshold suggests a connection with a potential correlation length that diverges at the transition. A first phenomenological interpretation was given by Wyart and coworkers [62], considering a rigid material and supposing that a blob of radius l is cut inside it. Keeping in mind the definition of the coordination number z, this operation yields a certain number of broken contacts on the surface, which is proportional to zl^{d-1}. Conversely, the number of excess contacts in the bulk is order $l^d \Delta z$. The smallest part that can be cut preventing the formation of floppy modes can be determined by equating the contributions of the bulk and of the surface, hence:

$$l^* \sim \frac{z}{\Delta z} \, . \tag{3.9}$$

The numerator can be safely assumed to be constant at jamming so that $l^* \sim 1/\Delta z$. One might wonder which is the most suitable way to cut contacts in order to create soft modes, namely a smooth function—having as many zeros as the number of broken contacts - should be introduced to modulate such modes. One should find a good compromise between the increasing number of soft/anomalous modes and the nodes of the modulating function. In other words, the plateau in the density of states is mostly due to modes with a characteristic length smaller than l^*. Conversely, for lengths greater than l^* the system is comparable to a continuous elastic medium.

To directly detect such a diverging length in practice, one can make use of the relation that links the crossover frequency to the wavelength λ. Then, from the analysis of transverse and longitudinal modes, one can observe two different behaviors, namely $\lambda_T \sim 1/\sqrt{\Delta z}$ and $\lambda_L \sim 1/\Delta z$. The scaling of the transverse length λ_T has been also confirmed in simulations [56], estimating the critical exponent that connects the correlation length and the distance from jamming as $\nu_T = 0.24 \pm 0.03$. Other possible scales can be introduced, even though their relation with $\lambda_{L,T}$ is not completely understood. Summarizing, the jamming transition encompasses two different facets, a discontinuous behavior in the coordination number and more diverging length scales, in line with the definition of a RFOT of Chap. 2.

3.3.5 Beyond the Spherical Symmetry

All the aspects we discussed in the previous Sections concern identical spherical particles. We wish to open now a brief parenthesis on more realistic models, going beyond the ideal assumption of frictionless spherical particles.

Thanks to isostaticity, at jamming the coordination number is exactly equal to twice the dimension of the system. However, this peculiar feature is correlated to the particle shape, changing for instance the isostatic number from 6 to 10 in the case of ellipsoids in three dimensions [40]. If spheres are deformed into more complex objects, one should also take into account rotational degrees of freedom, which in turn generate a new band of soft modes that lie below the band of translational modes (Fig. 3.7).

We discussed above the connection between a relevant correlation length at jamming and the crossover frequency ω^*, supposing to remove a blob from the system and to create from broken contacts new floppy modes. Considering each floppy mode distorted by a plane wave in such a way to satisfy some boundary conditions leads to a relation between ω^*, the mass of the particles and their stiffness [40], i.e.:

$$\omega^* \sim \sqrt{\frac{k_{\text{eff}}}{m}} (\phi - \phi_c)^{1/2} . \tag{3.10}$$

The resulting scaling for the onset frequency ω^* of the upper band of modes, obtained by deforming the shape of the particles by an amount ϵ or by compressing them, is actually the same, at least in three and two dimensions. This means that, while changing the particle shape entails new considerations, and a new band of modes, the scenario behind the jamming transition appears to be robust.

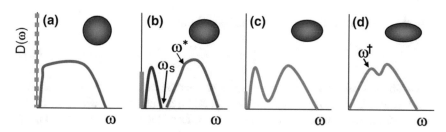

Fig. 3.7 Sketch of the density of states for frictionless soft ellipsoids, taken from [30]. The different colors refer to floppy modes (gray), rotational modes (blue), translational modes (red) and finally hybridized modes (green). For frictionless spheres only the translational band is considered, while upon deforming the particle shape also rotational degrees of freedom must be taken into account. The characteristic frequencies ω_s and ω^* depend on the degree of asphericity and on the variation $z - z_c^{sphere}$ respectively (in (b)). When the contact number $\to (z - z_c)^{ellip}$, the two bands merge in a unique hybridized band

3.4   Force and Pair Distributions at Random Close Packing

To investigate the inter-particle force distribution of dense amorphous packings let us take packings at ϕ_j and reduce their density. We can measure in principle the average momentum exchanged between two particles in contact over a time window $\tilde{t}$. In the very long-time regime $\tau \gg \tilde{t}$, the packing will relax towards more stable and compact structures, reaching either the values ϕ_{GCP} or ϕ_{FCC}. Anyway, we want to focus on intermediate time scales, attempting to reproduce the typical shape of $P(f)$, provided that forces are correctly normalized, i.e. $\langle f \rangle = 1$.

From the pair correlation function $g(r)$ of the spheres in the glass state, one can capture other interesting features. We shall indicate as $r = D$ the point at which the correlation function displays a delta peak. According to [19], the shape of $P(f)$ can be actually correlated to this delta peak, namely:

$$\frac{g(\lambda)}{g_G(D)} = \int_0^\infty df\, P(f) e^{-\lambda f} \,, \tag{3.11}$$

where $g_G(r)$ stands for the pair distribution in the glass phase, $\lambda = (r/D - 1)\, p$ and $p = \beta P/\rho$ for the reduced pressure. In particular one finds for $g_G(r)$:

$$g_G(r) \sim \begin{cases} \frac{1}{\phi_j - \phi} & r - D \sim \phi_j - \phi, \quad \lambda \sim 1 \\ \frac{1}{(\phi_j - \phi)\lambda^2} \sim \frac{\phi_j - \phi}{(r-D)^2} & \phi_j - \phi < r - D \sim \sqrt{\phi_j - \phi}, \quad \lambda \sim 1/\sqrt{\phi_j - \phi} \\ \exp\left[-(r-D)^2/(\phi_j - \phi)\right] & r - D \gg \sqrt{\phi_j - \phi}, \quad \lambda \gg 1/\sqrt{\phi_j - \phi} \end{cases} \tag{3.12}$$

and for the ratio:

$$\frac{g(\lambda)}{g_G(D)} = \frac{2}{\pi} \int_0^\infty df\, f e^{-f^2/\pi - \lambda f} \,. \tag{3.13}$$

In terms of $P(f)$ the above relation translates into:

$$P(f) \sim e^{-f^2/\pi} \,. \tag{3.14}$$

This scaling is valid for large forces and is in good agreement with numerical data. However, it turns out to deviate from the expected trend at small f, where the correct force distribution is related to $g_G(\lambda)$ at large λ. In this regime corrections to $g_G(\lambda)$ can no longer be neglected. The procedure applied before to derive the expressions at large f then fails. To correctly investigate the distribution of small forces, we have to take care of several effects. Let us see in detail a possibly reasonable description of this scenario.

In the introduction of this chapter we connected the jamming mechanism with the idea that isostatic packings are maximally random [58], which corresponds to the Edwards conjecture. Another line of thought is focused on the principle of marginal stability. However, by itself it cannot give a full explanation of non-linear processes, such as plasticity, thermal activation, etc. When only short-range interactions are present, a possible source of non-linearity is the creation or destruction of contacts between particles. In fact, these rearrangements do not constitute a continuous process but they occur intermittently via bursts or avalanches, whose size distribution is well fitted by a power law [17]. This kind of *crackling noise* [54] has been largely studied in different contexts, which go far beyond the physics of glasses, e.g. earthquakes. Wyart argued that a similar scenario occurs in amorphous packings of frictionless hard spheres [61], for which he derived a stability bound with respect to the opening and closing contacts. This bound actually interests the distributions of contact forces $P(f)$ and gaps between the particles $g(r)$. There is a general consensus on the typical trend of the pair distribution function $g(r)$ of hard spheres at RCP, confirmed in simulations and experiments. It exhibits a delta peak in $\sigma = 2r$, the contact distance, and the area below such a peak corresponds to the average coordination number z. Moreover, it is well-described by a power-law scaling, which can be related to a large number of touching particles as interpreted in a critical-phenomena perspective:

$$g(r) \sim (r - \sigma)^{-\gamma} \tag{3.15}$$

where the exponent $\gamma \approx 0.41$ [10, 13, 37]. For large values of the term in parenthesis, i.e. $r \to \infty$, a long-range order has been observed $g(r) - 1 \sim -r^{-4}$ [19]. This highlights a kind of similarity with the singular behavior of the static structure factor $S(k) \sim |k|$ for $k \to 0$ in long-ranged systems.

Similarly, the force distribution $P(f)$ has a peak at small values, in correspondence of the average value of f, and a power-law trend as $f \to 0^+$:

$$P(f) \sim f^\theta , \tag{3.16}$$

where the two exponents θ and γ are not independent, i.e. $\gamma \geq 1/(2 + \theta)$ [46, 61]. Extremely simple models and mean-field-like approximation predict the wrong value for these exponents $\theta = \gamma = 0$, which instead require a deep knowledge of all configurations visited by the dynamics. A full RSB calculation is thus needed [10, 23, 24]. As shown in Fig. 3.8, it was observed that, by considering a hard-sphere packing at jamming and opening contacts from it, both localized and extended modes appear [18, 38]. The main difference between these two modes is their decay with distance. The total force distribution can thus be written as a sum of these two contributions:

$$P(f) = n_l P_l(f) + (1 - n_l) P_{\text{ext}}(f) \tag{3.17}$$

where n_l decreases exponentially with the dimension d. For small f, the above distribution can be approximated with:

$$P(f) \approx n_l f^{\theta_l} + (1 - n_l) f^{\theta_{\text{ext}}} , \tag{3.18}$$

and hence, taking only localized modes into account, the exponent θ should be 0.17. To obtain the correct value of the exponent θ as predicted by the replica method, one

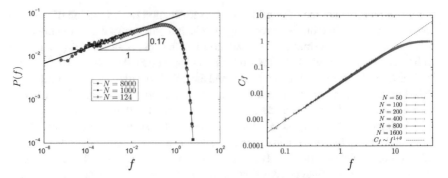

Fig. 3.8 When one considers a hard-sphere packing at the jamming density and opens contacts, both extended and localized modes are present. The second kind of modes gives rise to deformations that decay on short-range scales, order of a few grains. On the left the probability distribution $P(f)$, for isostatic soft spheres, with an exponent $\theta_1 \approx 0.17$. On the right the cumulative function of the force distribution $\int_0^f d\tilde{f} p(\tilde{f})$ for different numbers of particles N (in the perceptron model, the topic of next chapter). The straight line is $\sim f^{1+\theta}$, where θ can be exactly computed, i.e. $\theta \approx 0.42311$. The first exponent is obtained by considering localized modes, while the second value is in the case of extended modes. Figures taken from [18, 25]

should notice that the probability to have localized modes is exponentially small in high dimension and then $\theta = \theta_{\text{ext}} \approx 0.42$. As $\Delta\phi/\phi_c \to 0$ and $d \to \infty$ the localization of soft modes can thus be considered an irrelevant effect [14].

References

1. Anderson PW (1995) Through the glass lightly. Science 267:1615
2. Aste T, Coniglio A (2004) Cell theory for liquid solids and glasses: from local packing configurations to global complex behaviors. EPL (Europhys Lett) 67(2):165
3. Baule A et al (2018) Edwards statistical mechanics for jammed granular matter. Rev Modern Phys 90(1):015006
4. Berryman JG (1983) Random close packing of hard spheres and disks. Phys Rev A 27(2):1053
5. Biroli G, Mézard M (2001) Lattice glass models. Phys Rev Lett 88(2):025501
6. Buchenau U, Nücker N, Dianoux AJ (1984) Neutron scattering study of the low-frequency vibrations in vitreous silica. Phys Rev Lett 53(24):2316
7. Buchenau U et al (1986) Low-frequency modes in vitreous silica. Phys Rev B 34(8):5665
8. Castellani T, Cavagna A (2005) Spin glass theory for pedestrians. J Stat Mech: Theory Exp 2005(05):P05012
9. Charbonneau P et al (2014) Exact theory of dense amorphous hard spheres in high dimension. III. The full replica symmetry breaking solution. J Stat Mech: Theory Exp 2014(10):P10009
10. Charbonneau P et al (2014) Fractal free energy landscapes in structural glasses. Nat Commun 5:3725
11. Charbonneau P et al (2017) Glass and jamming transitions: from exact results to finite-dimensional descriptions. Ann Rev Condens Matter Phys 8:265
12. Charbonneau P et al (2015) Jamming criticality revealed by removing localized buckling excitations. Phys Rev Lett 114(12):125504

13. Charbonneau P et al (2012) Universal microstructure and mechanical stability of jammed packings. Phys Rev Lett 109(20):205501
14. Charbonneau P et al (2016) Universal non-Debye scaling in the density of states of amorphous solids. Phys Rev Lett 117(4):045503
15. Chen K et al (2011) Measurement of correlations between low-frequency vibrational modes and particle rearrangements in quasi-two-dimensional colloidal glasses. Phys Rev Lett 107(10):108301
16. Ciamarra MP et al (2003) Lattice glass model with no tendency to crystallize. Phys Rev E 67(5):057105
17. Combe G, Roux J (2000) Strain versus stress in a model granular material: a Devil's staircase. Phys Rev Lett 85(17):3628
18. DeGiuli E et al (2014) Force distribution affects vibrational properties in hardsphere glasses. Proc Natl Acad Sci 111(48):17054
19. Donev A, Torquato S, Stillinger FH (2005) Unexpected density fluctuations in jammed disordered sphere packings. Phys Rev Lett 95:090604
20. Durian DJ, Weitz DA (2005) Foam. In: Van Nostrand's encyclopedia of chemistry
21. Edwards SF (1991) The aging of glass forming liquids. Disord Condens Matter Phys 147
22. Edwards SF, Oakeshott RBS (1989) Theory of powders. Phys A Stat Mech Appl 157(3):1080
23. Franz S, Parisi G (2016) The simplest model of jamming. J Phys A Math Theor 49(14):145001
24. Franz S et al (2015) Universal spectrum of normal modes in low-temperature glasses. Proc Natl Acad Sci 112(47):14539
25. Franz S et al (2017) Universality of the SAT-UNSAT (jamming) threshold in nonconvex continuous constraint satisfaction problems. Sci Post Phys 2(3):019
26. Gardner E (1985) Spin glasses with p-spin interactions. Nucl Phys B 257:747
27. Grigera TS et al (2003) Phonon interpretation of the 'boson peak' in supercooled liquids. Nature 422(6929):289
28. Gross DJ, Kanter I, Sompolinsky H (1985) Mean-field theory of the Potts glass. Phys Rev Lett 55(3):304
29. Hansen IR, McDonald J-P (1990) Theory of simple liquids. Elsevier
30. van Hecke M (2009) Jamming of soft particles: geometry, mechanics, scaling and isostaticity. J Phys Condens Matter 22(3):033101
31. Ikeda A, Berthier L, Biroli G (2013) Dynamic criticality at the jamming transition. J Chem Phys 138:12A507
32. Itzykson C, Drouffe J-M (1991) Statistical field theory: volume 2, strong coupling, Monte Carlo methods, conformal field theory and random systems. vol 2. Cambridge University Press
33. Kamien RD, Liu AJ (2007) Why is random close packing reproducible? Phys Rev Lett 99(15):155501
34. Kurchan J, Maimbourg T, Zamponi F (2016) Statics and dynamics of infinite dimensional liquids and glasses: a parallel and compact derivation. J Stat Mech: Theory Exp 2016(3):033210
35. Kurchan J, Parisi G, Zamponi F (2012) Exact theory of dense amorphous hard spheres in high dimension. I. The free energy. J Stat Mech: Theory Exp 2012(10):P10012
36. Kurchan J et al (2013) Exact theory of dense amorphous hard spheres in high dimension. II. The high density regime and the gardner transition. J Phys Chem B 117:12979
37. Lerner E, Düring G, Wyart M (2013) Low-energy non-linear excitations in sphere packings. Soft Matter 9(34):8252
38. Lerner E, Düring G, Wyart M (2013) Simulations of driven over damped frictionless hard spheres. Comput Phys Commun 184(3):628
39. Liu AJ, Nagel SR (1998) Nonlinear dynamics: jamming is not just cool any more. Nature 396(6706):21
40. Liu AJ, Nagel SR (2010) The jamming transition and the marginally jammed solid. Ann Rev Condens Matter Phys 1(1):347
41. Lubachevsky BD, Stillinger FH (1990) Geometric properties of random disk packings. J Stat Phys 60(5–6):561
42. Majmudar TS et al (2007) Jamming transition in granular systems. Phys Rev Lett 98(5):058001

43. Martiniani S et al (2017) Numerical test of the Edwards conjecture shows that all packings are equally probable at jamming. Nat Phys 13(9):848
44. Masciovecchio C et al (1996) Observation of large momentum phonon like modes in glasses. Phys Rev Lett 76(18):3356
45. Maxwell JC (1864) On the calculation of the equilibrium and stiffness of frames. Lond Edinb Dublin Philos Mag J Sci 27(182):294
46. Müller M, Wyart M (2015) Marginal stability in structural, spin, and electron glasses. Ann Rev Condens Matter Phys 6:177
47. O'Hern CS et al (2003) Jamming at zero temperature and zero applied stress: the epitome of disorder. Phys Rev E 68(1):011306
48. Parisi G, Zamponi F (2010) Mean-field theory of hard sphere glasses and jamming. Rev Mod Phys 82(1):789
49. Percus JK, Yevick GJ (1958) Analysis of classical statistical mechanics by means of collective coordinates. Phys Rev 110(1):1
50. Phillips WA, Anderson AC (1981) Amorphous solids: low-temperature properties, vol 24. Springer
51. Press WH et al (1996) Numerical recipes in Fortran 90, vol 2. Cambridge University Press
52. Rivoire O et al (2004) Glass models on Bethe lattices. Eur Phys J B 37:55
53. Robles M, López de Haro M (2003) On the liquid-glass transition line in monatomic Lennard-Jones fluids. EPL (Europhys Lett) 62(1):56
54. Sethna JP, Dahmen KA, Myers CR (2001) Crackling noise. Nature 410(6825):242
55. Silbert LE, Liu AJ, Nagel SR (2009) Normal modes in model jammed systems in three dimensions. Phys Rev E 79(2):021308
56. Silbert LE, Liu AJ, Nagel SR (2005) Vibrations and diverging length scales near the unjamming transition. Phys Rev Lett 9(9):098301
57. Taraskin SN et al (2001) Origin of the boson peak in systems with lattice disorder. Phys Rev Lett 86(7):1255
58. Torquato S, Stillinger FH (2010) Jammed hard-particle packings: from Kepler to Bernal and beyond. Rev Mod Phys 82(3):2633
59. Torquato S, Truskett TM, Debenedetti PG (2000) Is random close packing of spheres well defined? Phys Rev Lett 84(10):2064–2067
60. Weitz DA (2004) Packing in the spheres. Science 303(5660):968
61. Wyart M (2012) Marginal stability constrains force and pair distributions at random close packing. Phys Rev Lett 109:125502
62. Wyart M (2005) On the rigidity of amorphous solids. Ann Phys 30:1
63. Wyart M, Nagel SR, Witten TA (2005) Geometric origin of excess low frequency vibrational modes in weakly connected amorphous solids. Europhys Lett 72(3):486
64. Wyart M et al (2005) Effects of compression on the vibrational modes of marginally jammed solids. Phys Rev E 72(5):051306
65. Xu N et al (2010) An harmonic and quasi-localized vibrations in jammed solids, Modes for mechanical failure. EPL (Europhys Lett) 90(5):56001
66. Xu N et al (2007) Excess vibrational modes and the boson peak in model glasses. Phys Rev Lett 98(17):175502
67. Zeravcic Z, van Saarloos W (2008) Nelson DR Localization behavior of vibrational modes in granular packings. EPL (Europhys Lett) 83(4):44001

Chapter 4
An Exactly Solvable Model: The Perceptron

As explained in Chap. 3, the jamming transition in hard spheres has been widely studied under both equilibrium and off-equilibrium initial conditions. In the infinite-dimensional limit one might nevertheless expect important simplifications. The goal of this chapter is to present a systematic formalism to derive the TAP free energy in jammed systems, taking a continuous constraint satisfaction problem, the perceptron, as the starting point. A detailed study of the spectrum of the Hessian matrix of the potential is also presented.

This chapter is essentially based on two related works developed in the last years [1, 2].

4.1 The Perceptron Model in Neural Networks

Before entering into the details of our work, we give a general introduction on the perceptron model, which we borrow from the theory of neural networks. For historical reasons, we shall first present the original formulation, which is the binary version proposed by Gardner [24, 25]. We will then illustrate the main techniques used to study the space of interactions.

The perceptron model is defined by a network made of $p = \alpha N$ patterns of N-bit words $\xi_i^\mu = \pm 1$ with $\mu = 1, ..., p$ and $i = 1, ..., N$. The main purpose of this model is to capture network learning. To achieve this, the patterns μ must be the fixed points of a dynamical rule that governs the system's evolution. Generally one can consider a zero-temperature Monte Carlo dynamics given by the sign-function $\xi_0^\mu = \text{sgn}(\sum_{k=1}^N J_{0k}\xi_k^\mu)$. The other condition—to be satisfied at each site i and each pattern μ—that guarantees a finite basin of attraction, is outlined below:

$$\xi_i^\mu \left(h_i(\{\xi_j^\mu\}) - T_i \right) > \sigma \tag{4.1}$$

where σ is a positive parameter and T_i is a threshold, which depends on site i. For larger values of σ one should observe larger basins of attraction. The quantity h_i is the internal magnetic field, defined as:

$$h_i(\{\xi_j^\mu\}) = \frac{1}{\sqrt{N}} \sum_{j \neq i} J_{ij} \xi_j^\mu \tag{4.2}$$

where the couplings J_{ij} are chosen in such a way to have $\sum_{j \neq i} J_{ij}^2 = N$ for each i.

They are *annealed* variables and have only to satisfy a constraint on the sum of their squared values. On the other hand, the role of the couplings is to guarantee that the patterns ξ_i^μ are stored as fixed points of a given dynamics defined at the beginning.[1] A fundamental quantity in the problem is the fractional volume of the space of solutions occupied by these interactions, corresponding to the maximum storage capacity of the network. If T_i is set to zero, the volume $\alpha(\sigma)$ can be easily computed and its critical point falls at $\alpha = p/N = 2$ and $\sigma = 0$ [24]. The computation can be generalized for a generic threshold: however, if the constraints defined by Eq. (4.1) are correlated, a larger number of patterns can be stored at the cost of a smaller information capacity of the network.

To compute the space volume one needs to introduce the replica method [36]. Hence, we introduce the following shorthand notation, in the same spirit of the definition of the magnetization and the Edwards-Anderson order parameter in spin-glass literature:

$$m_i^\alpha = \frac{1}{\sqrt{N}} \sum_{j \neq i} J_{ij}^\alpha$$
$$q_i^{\alpha\beta} = \frac{1}{N} \sum_{j \neq i} J_{ij}^\alpha J_{ij}^\beta \qquad \alpha \neq \beta . \tag{4.3}$$

A replica-symmetric ansatz is considered, based on the assumptions:

$$m_i^\alpha = m , \qquad q_i^{\alpha\beta} = q . \tag{4.4}$$

As the parameter α increases, the overlap q increases as well as a consequence of the fact that different solutions are correlated. When $q \to 1$, the space of solutions shrinks to zero, determining the critical value $\alpha_c(\sigma)$. The volume fraction, satisfying the normalization condition for the J_{ij}'s and constrained over the patterns, reads [24]:

$$V = \frac{\int \prod_{j \neq i} dJ_{ij} \prod_{\mu, i} \theta(\xi_i^\mu h_i(\{\xi_j^\mu\}) - \sigma) \delta(\sum_{j \neq i} J_{ij}^2 - N)}{\int \prod_{i \neq j} dJ_{ij} \prod_i \delta(\sum_{j \neq i} J_{ij}^2 - N)} \tag{4.5}$$

---

[1] In replica literature, the patterns are quenched variables and the interaction strengths are annealed variables.

which can be rewritten using the factorization property on i: $V = \prod_{i=1}^{N} V_i$. In order to obtain the thermodynamics of this system, we need to compute the logarithm of the volume of solutions and average over the quenched distribution of patterns. This can be achieved exploiting the so-called replica trick, i.e.:

$$\overline{\log V} = \lim_{n \to 0} \frac{\overline{V^n} - 1}{n} \qquad (4.6)$$

Note that, as usual in the replica context, we are implicitly assuming the analytical continuation $n \to 0$ to exist. Then, the expression for the replicated volume fraction is given by:

$$\overline{V^n} = \frac{\prod_{a=1}^{n} \int \prod_{j \neq i} dJ_{ij}^a \prod_{\mu,i} \theta(\xi_i^\mu \sum_{j \neq i} \frac{J_{ij}^a}{\sqrt{N}} \xi_j^\mu - \sigma) \delta(\sum_{j \neq i} (J_{ij}^a)^2 - N)}{\prod_{a=1}^{n} \int \prod_{i \neq j} dJ_{ij}^a \prod_i \delta(\sum_{j \neq i} (J_{ij}^a)^2 - N)} \qquad (4.7)$$

where we introduced the replica index $a = 1, ..., n$. At this point we use the integral representation of the θ-function:

$$\theta\left(\xi_i^\mu \sum_{j \neq i} \frac{J_{ij}^a}{\sqrt{N}} \xi_j^\mu - \sigma\right) = \int_\sigma^\infty \frac{d\lambda_\mu^a}{2\pi} \int_{-\infty}^\infty dx_\mu^a \exp\left[ix_\mu^a\left(\lambda_\mu^a - \xi_i^\mu \sum_{j \neq i} \frac{J_{ij}^a}{\sqrt{N}} \xi_j^\mu\right)\right] \qquad (4.8)$$

to be plugged into Eq. (4.7). Since the patterns are binary variables the average over ξ_j^μ (for $j \neq i$) gives:

$$\exp\left[\sum_\mu \sum_{j \neq i} \ln \cos\left(\sum_a \frac{x_\mu^a J_{ij}^a}{\sqrt{N}}\right)\right], \qquad (4.9)$$

where we can use the asymptotic expansion of the term in square brackets[2] and rewrite the expression above as:

$$\exp\left[-\frac{1}{2} \sum_\mu \sum_{a,b} x_\mu^a x_\mu^b \left(\sum_{j \neq i} \frac{J_{ij}^a J_{ij}^b}{N}\right)\right]. \qquad (4.10)$$

Let us introduce the auxiliary variable q^{ab}:

$$q^{ab} = \frac{1}{N} \sum_{j \neq i} J_{ij}^a J_{ij}^b \qquad a < b, \qquad (4.11)$$

---

[2] We made use of the following expansion for the logarithm: $\ln(\cos(x))|_{x=0} \approx -\frac{1}{2}x^2 - \frac{1}{12}x^4 + o(x^4)$.

which is beneficial to rewrite the partition function as:

$$
\overline{V^n} = \frac{1}{Z} \int \prod_{a=1}^{n} dE^a \int \prod_{a<b} \frac{dq^{ab} d\hat{q}^{ab}}{2\pi/N} \exp\left[N\left(\alpha G(q^{ab}) + F(\hat{q}^{ab}, E^a) - \sum_{a<b} \hat{q}^{ab} q^{ab} + \frac{1}{2} \sum_a E^a \right) \right]
$$
(4.12)

where Z stands for the normalization, while $\hat{q}^{ab}$ and E^a are Lagrange multipliers associated with q^{ab} and with the θ-function constraint. The two quantities $G(q^{ab})$ and $F(\hat{q}^{ab}, E^a)$ read:

$$
G(q^{ab}) = \ln \prod_{a=1}^{n} \int_{-\infty}^{\infty} dx^a \int_{\sigma}^{\infty} \frac{d\lambda^a}{2\pi} \exp\left(i \sum_a x^a \lambda^a - \frac{1}{2} \sum_a (x^a)^2 - \sum_{a<b} q^{ab} x^a x^b \right) ,
$$

$$
F(\hat{q}^{ab}, E^a) = \ln \prod_{a=1}^{n} \int dJ^a \exp\left(-\frac{1}{2} \sum_a E^a (J^a)^2 + \sum_{a<b} \hat{q}^{ab} J^a J^b \right) .
$$
(4.13)

In the large-N limit we can safely perform a saddle-point approximation in the replica-symmetric (RS) ansatz [36]. Since the space of solutions is totally connected, the saddle-point method can be reasonably applied and leads to:

$$
q^{ab} = q , \qquad \hat{q}^{ab} = \hat{q} , \qquad E^a = E
$$
(4.14)

where the first two variables are considered for $a < b$, while the last one for all possible values of the replica index a. In the large-N limit and $n \to 0$, the resulting expression for the volume fraction turns out to be:

$$
\overline{V^n} = \exp\left[Nn \left(\min_q G(q) + O(1/N) \right) \right]
$$
(4.15)

where $G(q)$ reads:

$$
G(q) = \alpha \int \mathcal{D}t \ln H\left(\frac{\sqrt{q}t + \sigma}{\sqrt{1-q}} \right) + \frac{1}{2} \ln(1-q) + \frac{1}{2} \frac{q}{1-q}
$$
(4.16)

and $H(x) \equiv \int_x^{\infty} \mathcal{D}z$ represents the complementary error function.[3] The saddle-point equation for $G(q)$ with respect to q allows to derive the generic expression for the overlap, whose maximum value corresponds to the maximum storage capacity, i.e.:

$$
\alpha_c(\sigma) = \left(\int_{-\sigma}^{\infty} \mathcal{D}t (t+\sigma)^2 \right)^{-1} .
$$
(4.17)

As mentioned before, as $\sigma \to 0$, the critical value $\alpha_c(0) = 2$ can be immediately obtained [12, 24, 40].

---

[3] The calligraphic symbol stands for the Gaussian integration over t with zero mean and unit variance.

4.2 From Computer Science to Sphere-Packing Transitions

Several properties of the jamming transition—as the emergence of a power-law behavior in the distribution of the forces and gaps between the particles, the nature and the shape of vibrational modes—can be investigated taking the perceptron as the describing model. In fact, the problem of packing spheres in space can be regarded as a constraint satisfaction problems (CSPs): in a jammed system the particle motion is hindered by neighboring particles, which induce geometrical and mechanical constraints in terms of force and torque balance. The connection between jammed systems and the CSP paradigm, at the cornerstone of the theory of computational complexity, has been proposed in several works [6, 31, 45, 46]. However, finding a solution for a jamming-satisfaction problem is a challenging issue, which requires reasonable approximations in order to decouple the geometrical problem (e.g. in the determination of the contact network) from the mechanical one (to get, for instance, the full-force distribution). A possible way to achieve this goal might consist of fixing the geometry of the packing on a random graph and then in computing the force distribution by the *cavity method* [34, 36]. This force distribution is nothing but the Edwards measure introduced in Chap. 3, Θ_J, over all the solutions satisfying the required constraints for a given realization of the network [4]. If the partition function defined over the uniform probability distribution is greater or equal than one, a solution does exist and the problem is satisfiable (SAT). In the opposite case, no solution can satisfy all the constraints (UNSAT). Actually one can distinguish a region where the problem is undetermined from a region where it is overdetermined. Several studies have been done especially on random graphs, where a solution was exactly found. Nevertheless, further developments in this field have been made possible once it was realized that sphere systems in high dimension belong to the same universality class as the continuous perceptron, according to a new interpretation proposed by Franz and Parisi [21] considering a spherical perceptron. The continuous nature of the variables adds here a new dimension to the problem making the SAT-UNSAT transition exactly the same as the equilibrium jamming transition.

In this thesis we propose a modified version of the perceptron model *à la Gardner* with a particular emphasis on a non-convex regime responsible for non-trivial properties in the space of solutions [2, 22, 23]. This simple model has the advantage of allowing a much more direct derivation of the fullRSB solution than in hard-sphere glasses [37]. In the following, we shall first introduce the Franz–Parisi model and then we shall present our main results in the SAT phase, which mainly concern the derivation of an effective mean-field-like potential and the study of vibrational modes as well as relevant scaling properties near jamming.

The Franz–Parisi toy model is a remarkable starting point for studying glassy phenomena and jamming in high dimension, consisting of M obstacles randomly distributed over a spherical surface in N dimensions, with radius $\sqrt{N}$. The Hamiltonian of the model depends on $M = \alpha N$ random *gaps* $h_\mu(\vec{x})$ (where $\mu = 1, ..., M$) via a soft-constraint interaction:

$$\mathcal{H}[\vec{x}] = \frac{1}{2} \sum_{\mu=1}^{M} h_{\mu}^2(\vec{x}) \theta(-h_{\mu}(\vec{x})) \,, \tag{4.18}$$

where $\theta(x)$ is the Heaviside function. Therefore, one can associate this setting to an optimization problem, whose *cost function* is finite if at least one constraint is not satisfied and zero otherwise. The choice of Eq. (4.18) is arbitrary, only motivated by a clear analogy with harmonic soft spheres. The gaps are functions of the system configuration $\vec{x} = \{x_1, ..., x_N\}$, defined on a N-dimensional hypersphere, i.e. $\sum_{i=1}^{N} x_i^2 = N$, and they satisfy the following relation:

$$h_{\mu}(\vec{x}) = \sum_{i=1}^{N} \frac{\xi_i^{\mu} x_i}{\sqrt{N}} - \sigma \,, \qquad \forall \mu = 1, .., M \,. \tag{4.19}$$

This corresponds to the scalar product between the random obstacles ξ_i^{μ} ($i = 1, ..., N$ and $\mu = 1, ..., M$) and the reference particle position, both defined on the sphere. The components ξ_i^{μ}, which play the role of quenched disorder, are i.i.d. Gaussian random variables with zero mean and unitary variance, i.e. $\mathcal{N}(0, 1)$. The scaling $M = \alpha N$ is chosen to guarantee a well-defined thermodynamic limit, with which the thermodynamic analysis will be concerned.

Depending on the positive or negative value of the control parameter σ, two different situations can occur: for positive σ the model defines the usual perceptron classifier, which gives rise to a convex optimization problem (Fig. 4.1), whereas for negative values of σ the space of solutions is no longer convex, losing its ergodicity properties and inducing anomalous features. This is the novelty with respect to the original model proposed by Gardner. Indeed, the second scenario can be mapped into the problem of a single dynamical sphere in a background of *quenched* obstacles ξ^{μ}, called *patterns* in neural network terminology. This is exactly the regime we are interested in. For this purpose, it might be helpful to define a dictionary to connect this model to particle systems: for instance, the energy is proportional to $[h^2]$ while the pressure, defined as $\partial H/\partial \sigma$, is proportional to $[h]$ in both cases. Once the constraint μ is defined, we can directly give the expression of the forces, acting from μ to a variable i, i.e. $f_i^{\mu} = -(\partial h_{\mu}/\partial x_i) h_{\mu} \theta(-h_{\mu})$. They correspond to the single contribution of the total force $F_i = -\partial H/\partial x_i$ on the particle x_i.

By conveniently varying the two tunable parameters, σ and $\alpha = M/N$, one might explore different regions of the phase diagram. These two parameters control the transition analogously to the packing fraction in sphere problems. In particular, the system might undergo a critical transition determining the passage from a satisfiable region, SAT phase, to an unsatisfiable one, UNSAT phase. The first one corresponds to a hard-sphere regime defined by a zero energy manifold (in agreement with all the constraints $h_{\mu}(x) > 0$), whereas the second scenario can be mapped to a soft-sphere problem described by a harmonic potential in $h_{\mu}(\vec{x})$ (see Eq. (4.18)). Physically, this SAT/UNSAT transition coincides with the jamming transition in correspondence

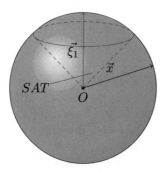

Fig. 4.1 Representation of the perceptron model in the convex case, i.e. for $\sigma > 0$, with only one constraint. The scalar product between $\vec{x}$ and $\vec{\xi_1}$ should be bigger than a threshold, excluding all the solutions in the region below. The solution space is now convex. Considering more constraints the space of solutions reduces but remains convex. Both the reference particle and the obstacles live on a N-dimensional sphere. For $\sigma < 0$ the solution space would be given by the intersection of non-convex domains

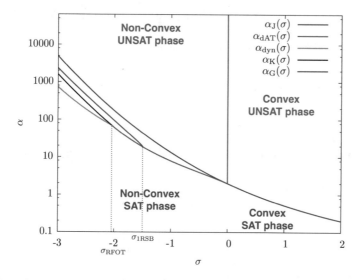

Fig. 4.2 Perceptron phase diagram, reprinted from [23], showing the control parameter $\alpha = M/N$ as a function of σ. For $\sigma > 0$ the perceptron defines a convex problem, while for $\sigma < 0$ new critical properties emerge. In this second regime the perceptron can be exactly mapped to sphere problems. The SAT (hard-sphere) regime has at least one configuration that satisfies all the constraints $h_\mu(\vec{x}) > 0$, while the UNSAT (soft-sphere) regime has no such configuration

of which the volume of the space of solutions satisfying the given assignments continuously shrinks to zero.

An important quantity to pay attention to is the Gardner volume [24] of the satisfying assignments:

$$\mathcal{V}(\alpha, \sigma) \equiv e^{NS(\alpha,\sigma)} = \int_{S_N} d\vec{x} \prod_{\mu=1}^{\alpha N} \theta(h_\mu) \,, \tag{4.20}$$

where S_N stands for the N-dimensional hypersphere of radius $\sqrt{N}$. A vanishing value of the volume $\mathcal{V}(\alpha, \sigma)$ signals the approach of the jamming transition. Using the replica method one can compute the quenched average of the entropy $S(\alpha, \sigma)$ over the random variables ξ^μ and show that it can be expressed as a saddle-point computation over the overlap matrix Q_{ab}, with $a, b = 1, ..., n$ in the $n \to 0$ limit.

To better investigate the interesting features of the model one should look at the zero-temperature limit of the partition function, then at the free energy. In general the intensive free energy f is given by $f = e - Ts$, where e denotes the energy per particle and s the thermodynamic entropy. In the zero-temperature limit, in the SAT phase the energy tends to zero: this should be clear by looking at Eq. (4.18). Therefore, the only contribution to the free energy comes from the finite entropy term: as a consequence $-\beta f = s$ in the SAT phase. The UNSAT phase is instead characterized by a finite-energy contribution, which actually coincides with the free energy since $Ts \to 0$ [23].

The partition function of this model is defined as:

$$Z = \int \mathcal{D}\vec{x} e^{-\beta H[\vec{x}]} = \int \mathcal{D}\vec{x} \prod_{\mu=1}^{M} \left[dr_\mu e^{-\beta v(r_\mu - \sigma)} \delta \left(r_\mu - \frac{1}{\sqrt{N}} \vec{x} \cdot \vec{\xi}^\mu \right) \right] \tag{4.21}$$

where $\mathcal{D}\vec{x}$ is the uniform measure over the N-dimensional sphere and $v(h)$ is a generic potential that depends on the gap variables. It will be better defined in the following.

In order to compute the average free energy, we need to introduce replicas and to consider the following relation, equivalent to Eq. (4.6):

$$f = -\frac{1}{\beta N} \overline{\log Z} = -\frac{1}{\beta N} \lim_{n \to 0} \partial_n \overline{Z^n} \,. \tag{4.22}$$

To write the replicated partition function we constrain the value of the random gaps and we introduce the variables $\hat{r}_\mu$ conjugated to the gaps themselves:

$$\overline{Z^n} \propto \int \left(\prod_{a=1}^{n} \mathcal{D}\vec{x}^a \right) \left(\prod_{a=1}^{n} \prod_{\mu=1}^{M} dr_\mu^a d\hat{r}_\mu^a \right) e^{\sum_{\mu,a} i\hat{r}_\mu^a \left(r_\mu^a - \frac{1}{\sqrt{N}} \vec{x}^a \cdot \vec{\xi}_\mu \right)} \prod_{a=1}^{n} \prod_{\mu=1}^{M} e^{-\beta v(r_\mu^a - \sigma)} \,. \tag{4.23}$$

The average over the quenched disorder, given a Gaussian distribution for the patterns $\vec{\xi}_\mu$, leads to:

$$\overline{e^{-\frac{1}{\sqrt{N}} \sum_{\mu,a} i\hat{r}_\mu^a \vec{x}^a \cdot \vec{\xi}_\mu}} = e^{-\frac{1}{2} \sum_{a,b,\mu} \hat{r}_\mu^a \hat{r}_\mu^b Q^{ab}} \tag{4.24}$$

where a $n \times n$ overlap matrix has been introduced, i.e. $Q^{ab} \equiv \frac{1}{N} \vec{x}^a \cdot \vec{x}^b$. The product over μ is factorized and yields:

$$
\overline{Z^n} \propto \int \prod_{a=1}^{n} \mathcal{D}\vec{x}^a \left[\int \left(\prod_{a=1}^{n} dr^a d\hat{r}^a \right) e^{\sum_a i\hat{r}^a r^a - \frac{1}{2} \sum_{ab} \hat{r}^a \hat{r}^b Q^{ab} - \beta \sum_a v(r^a - \sigma)} \right]^M
$$

$$
\propto \int \prod_{ab} dQ^{ab} e^{\frac{N}{2} \log \det Q} \left[\int \left(\prod_{a=1}^{n} dr^a d\hat{r}^a \right) e^{\sum_a i\hat{r}^a r^a - \frac{1}{2} \sum_{ab} \hat{r}^a \hat{r}^b Q^{ab} - \beta \sum_a v(r^a - \sigma)} \right]^M .
$$

(4.25)

The last line has been obtained by a variable change, from $\vec{x}^a$ to Q^{ab}, where $e^{\frac{N}{2} \log \det Q}$ is the Jacobian of the transformation. By performing a Taylor expansion of the exponential and using the Wick theorem, the quadratic term can be rewritten in the following way:

$$
\int \mathcal{D}\varrho \vec{k} f \left(\{k^a\} \right) = e^{\frac{1}{2} \sum_{ab} Q^{ab} \frac{\partial^2}{\partial k^a \partial k^b}} f \left(\{k^a\} \right) \Big|_{k^a = 0}
$$

(4.26)

$$
e^{-\frac{1}{2} \sum_{ab} \hat{r}^a \hat{r}^b Q^{ab}} = e^{\frac{1}{2} \sum_{ab} Q^{ab} \frac{\partial^2}{\partial k^a \partial k^b}} e^{-\sum_a k^a i\hat{r}^a} \Big|_{k^a = 0} .
$$

(4.27)

This leads to further simplifications of Eq. (4.25):

$$
\overline{Z^n} \propto \int \prod_{ab} dQ^{ab} e^{\frac{N}{2} \log \det Q} \left[\int \left(\prod_{a=1}^{n} dr^a d\hat{r}^a \right) e^{\sum_a i\hat{r}^a (r^a - k^a) + \frac{1}{2} \sum_{ab} Q^{ab} \frac{\partial^2}{\partial h^a \partial h^b} - \beta \sum_a v(r^a - \sigma)} \Big|_{k^a = 0} \right]^M
$$

$$
\propto \int dQ e^{\frac{N}{2} \log \det Q + M \log \left(e^{\frac{1}{2} \sum_{ab} Q^{ab} \frac{\partial^2}{\partial h^a \partial h^b}} \prod_a e^{-\beta v(h^a)} \Big|_{h^a = -\sigma} \right)}
$$

(4.28)

where we notice that the integration over $\hat{r}^a$ gives $\delta(r^a - k^a)$. Hence, the partition function can be rewritten in shorthand notation as:

$$
\overline{Z^n} = \int dQ e^{NS(Q)} ,
$$

(4.29)

$$
S(Q) = \frac{1}{2} \log \det Q + \alpha \log \left(e^{\frac{1}{2} \sum_{ab} Q_{ab} \frac{\partial^2}{\partial h_a \partial h_b}} \prod_a e^{-\beta v(h_a)} \Big|_{h_a = -\sigma} \right)
$$

(4.30)

where we remind that $\alpha = M/N$. Once the generic expression for the replicated free energy is known, we can concentrate our efforts on the analysis of the typical behavior in both the SAT and the UNSAT phase. In particular, at the RS level the free energy reads:

$$\lim_{n \to 0} \frac{1}{n} \left[\frac{1}{2} \log \det Q + \alpha \log \left(e^{\frac{1}{2} \sum_{ab} Q_{ab} \frac{\partial^2}{\partial h_a \partial h_b}} \prod_a e^{-\beta v(h_a)} \bigg|_{h_a = -\sigma} \right) \right] =$$

$$= \frac{1}{2} \left[\log(1 - q) + \frac{q}{1 - q} \right] + \alpha \gamma_q * f(q, h)|_{h_a = -\sigma} \tag{4.31}$$

where the symbol $*$ stands for the convolution with a Gaussian distribution, namely:

$$\gamma_q * f(h) = \int_{-\infty}^{\infty} \mathcal{D}t_q f(h - t) . \tag{4.32}$$

and $\mathcal{D}t_q$ stands for the centred Gaussian measure with variance q. The function $f(q, h)$ verifies the Parisi equation as it will be clarified later on.

4.2.1 Free-Energy Behavior in the SAT Phase

To highlight the differences between the SAT and the UNSAT phases, one should first notice that in the former, thanks to the finiteness of the phase-space volume, two typical solutions can be in different states and $q < 1$. Conversely, in the UNSAT phase there is one single minimum and as $T \to 0$ all solutions converge to this, which is equivalent to state that $q \to 1$. Let us focus before on the SAT (hard-sphere) phase, where the gaps can take only positive values. In this regime the second term of the free energy reduces to [23]:

$$f(q, h)_{\text{SAT}} = \lim_{\beta \to \infty} \log \gamma_{1-q} * e^{-\beta v(h)} \equiv \log \gamma_{1-q} * \theta(h) =$$

$$= \log \int_{-\infty}^{h} \frac{dt}{\sqrt{2\pi(1-q)}} e^{-\frac{t^2}{2(1-q)}} \equiv \log \left\{ \frac{1}{2} \left[1 + \text{erf} \left(\frac{h}{\sqrt{2(1-q)}} \right) \right] \right\} . \tag{4.33}$$

4.2.2 Free-Energy Behavior in the UNSAT Phase

In the UNSAT phase the only possibility is $q = 1$ because of the presence of a unique minimum in the energy landscape. The harmonic vibrations around this minimum can be well approximated by $q = 1 - \chi T + O(T^2)$, where the parameter χ diverges at the jamming transition. The second term of the free energy reads in this case [23]:

$$f(q, h)_{\text{UNSAT}} = \log \gamma_{1-q} * e^{-\beta v(h)} \equiv \log \gamma_{1-q} * e^{-\beta \frac{h^2}{2} \theta(-h)} . \tag{4.34}$$

With a marginal change of notation, this can be rewritten as [23]:

$$f(q,h)_{\text{UNSAT}} = \log \int_{-\infty}^{+\infty} \frac{dt}{\sqrt{2\pi\chi T}} e^{-\frac{(h-t)^2}{2\chi T} - \frac{t^2}{2T}\theta(-t)} . \tag{4.35}$$

In the zero-temperature limit and for $h \sim O(1)$, the solutions of the above integral can be found using the saddle-point approximation, i.e.:

$$\frac{t-h}{\chi} + t\theta(-t) = 0 , \tag{4.36}$$

which implies:

$$t^*(h) = h \quad \text{if } h > 0 , \qquad t^*(h) = \frac{h}{1+\chi} \quad \text{if } h < 0 . \tag{4.37}$$

Keeping that in mind, we can rewrite the solution for the free energy as:

$$f(q,h)_{\text{UNSAT}} = \log\left[\theta(h) \int_{-\infty}^{+\infty} \frac{dy}{\sqrt{2\pi\chi T}} e^{-\frac{y^2}{2\chi T}} + \theta(-h) e^{-\frac{h^2}{2(1+\chi)T}} \int_{-\infty}^{\infty} \frac{dy}{\sqrt{2\pi\chi T}} e^{-\frac{1+\chi}{2\chi T}y^2} \right] \tag{4.38}$$

where we have defined $y = t - t^*(h)$. Neglecting subleading logarithmic terms, we finally get:

$$f(1 - \chi/\beta, h)_{\text{UNSAT}} \approx -\frac{\beta h^2}{2(1+\chi)}\theta(-h) . \tag{4.39}$$

Putting all these results together, the zero-temperature free energy (4.31) becomes:

$$\lim_{\beta\to\infty} f_{\text{UNSAT}} = -\frac{1}{2\chi} + \frac{\alpha}{2(1+\chi)} \int_{-\infty}^{\sigma} \frac{dh}{\sqrt{2\pi}} e^{-\frac{h^2}{2}}(h-\sigma)^2 , \tag{4.40}$$

which, differentiated with respect to χ, provides a fundamental relation, valid in the UNSAT phase (for $\alpha > \alpha_c(\sigma)$), which links χ to the critical capacity:

$$\left(1 + \frac{1}{\chi}\right)^2 = \frac{\alpha}{\alpha_c(\sigma)} . \tag{4.41}$$

The jamming limit is identified by the condition $\alpha \to \alpha_c(\sigma)$, which in turn implies a divergent χ.

4.2.3 Jamming Regime

The jamming line lies in the fullRSB phase, where the entropy has to be derived under the condition to parametrize the matrix Q in terms of the function $q(x)$ defined in the interval $[q_0, q_1]$. The inverse function $x(q)$ is parametrized over the interval $[q_0, q_1]$.

Then, the first term in the expression for the entropy (4.30) reads [21]:

$$\frac{1}{n} \operatorname{tr} \log Q = \log(1 - q_1) + \frac{q_0}{\lambda(q_0)} + \int_{q_0}^{q_1} dq \frac{1}{\lambda(q)} \,,$$

$$\lambda(q) = 1 - q_1 + \int_q^{q_1} dq' x(q') \tag{4.42}$$

while the second one can be rewritten as $-n\alpha \int \mathcal{D}_{q_0}(h - \sigma) f(q_0, h)$, where the function $f(q, h)$ satisfies the following differential equation [36]:

$$\frac{\partial f}{\partial q} = -\frac{1}{2}\left[\frac{\partial^2 f}{\partial h^2} + x \left(\frac{\partial f}{\partial h} \right)^2 \right] \tag{4.43}$$

with the boundary condition:

$$f(q_1, h) = -\log H \left(\frac{\sigma - h}{\sqrt{1 - q_1}} \right) . \tag{4.44}$$

The distribution of the local gaps at level q, i.e. $P(q, h)$, verifies the following equations:

$$\frac{\partial P(q, h)}{\partial q} = \frac{1}{2} \left(\frac{\partial^2 P(q, h)}{\partial h^2} - 2x \frac{\partial (m P(q, h))}{\partial h} \right) , \qquad P(q_0, h) = D_{q_0}(h)/dh \tag{4.45}$$

where $m(q, h) \equiv \partial f(q, h)/\partial h$ and satisfies a similar equation:

$$\frac{\partial m(q, h)}{\partial q} = -\frac{1}{2} \left(\frac{\partial^2 m(q, h)}{\partial h^2} + 2xm \frac{\partial m(q, h)}{\partial h} \right) . \tag{4.46}$$

Considering the variational equations with respect to $x(q)$ one can get:

$$\frac{1}{2} \left(\frac{q_0}{\lambda(q_0)^2} + \int_{q_0}^q dq' \frac{1}{\lambda(q)^2} \right) - \frac{\alpha}{2} \int dh P m^2 = 0 \,, \tag{4.47}$$

which is the starting point to evaluate the system's stability. An important quantity to define the stability is the *replicon eigenvalue*, which comes out as soon as one considers the derivative of the previous expression with respect to q, i.e.:

$$\frac{1}{2\lambda(q)^2} - \frac{\alpha}{2} \int dh P m'^2 = 0 \,. \tag{4.48}$$

A stable solution requires the l.h.s. to be positive. Conversely, the vanishing value of the replicon mode signals the marginal stability of the RSB solution, associated with a divergent spin glass susceptibility, as usual in a continuous transition. As shown in Fig. 4.2, for positive values of σ the solution can be only replica-symmetric, while for negative values a RSB transition appears that can occur either via a second order transition to a continuous one or via a discontinuous RFOT. From $\sigma = 0$ to $\sigma_{1RSB} \approx -2.05$ [23] one finds a de Almeida-Thouless instability line [15, 36], below which a one-step transition appears. However, approaching the jamming line one should expect a continuous transition. Hence, in the perceptron one could identify two phases, a hypostatic regime for $\sigma > 0$ and an isostatic and marginally stable regime for $\sigma < 0$. Using Eqs. (4.45) and (4.46), one can study the scaling regime close to jamming, namely as $(1 - q_1)$ tends to zero. By analyzing their behavior for large values of $|h|$ and properly matching the two regimes, respectively for large positive and large negative values of the gaps, one finally gets in the jamming limit:

$$P(q, h) \sim A\delta(h) + p(h) \tag{4.49}$$

where $A \equiv \int du\, p_0(u)$ is given by the normalization condition, while $p(h)$ represents the true physical distribution of the gaps at jamming, scaling as a power law as $h \to 0$, i.e. $p(h) \sim h^{-\gamma}$ with an exponent $\gamma = 0.41269$. The relation between the critical exponents governing the decay in the distribution of the gaps and the forces has been first conjectured by Wyart in [41]. This principle can be reformulated in a few passages in the perceptron model, by looking at the dual problem of finding the maximum value of the $\vec{x}^2$ in the free space $\mathbb{R}^N$ at fixed negative σ. Once the Lagrangian of the model is defined, one can analyze its stability by opening a single contact and hence obtain the same estimation for $\gamma \geq \frac{1}{2+\theta}$ [21].

4.3 Computation of the Effective Potential in Fully-Connected Models

Several questions concerning glasses in the low-temperature regime still need an answer. In this light, the possibility of establishing a unifying framework for jamming, irrespective of microscopic details and specific settings, looks very intriguing. We aim at determining an effective thermodynamic potential that might properly describe the perceptron model and, in a broader perspective, that can capture interesting features shared by different jammed systems.

To let the reader fully understand our analysis, we first introduce the computation of the free energy functional with a special emphasis on fully-connected (FC) structures. Even in a disordered system, for a given realization of the disorder, each state can be identified either in terms of the magnetization m or of the density profile ρ, although in many cases this characterization could be too approximate [44]. For simplicity, we consider a spin system identified by the site variables σ_i and by external magnetic fields b_i. The extensive free energy then reads:

$$\Omega[\vec{b}] = -T \log \sum_{\vec{\sigma}} e^{-\beta H[\vec{\sigma}] + \beta \sum_i b_i \sigma_i} . \tag{4.50}$$

Essentially we want to constrain the system to have a given magnetization $\vec{m}$ and to do this we need to introduce auxiliary local magnetic fields. Starting from the definition (4.50) and taking its Legendre transform, we obtain the corresponding expression in terms of conjugated variables $\vec{m}$:

$$-\beta F[\vec{m}] = -\beta \max_{\vec{b}} \left[\Omega[\vec{b}] + \sum_i b_i m_i \right] = \min_{\vec{b}} \left[\log \sum_{\vec{\sigma}} e^{-\beta H[\vec{\sigma}] + \beta \sum_i b_i (\sigma_i - m_i)} \right]. \tag{4.51}$$

From this, one can immediately derive the following relations:

$$\frac{\partial F[\vec{m}]}{\partial m_i} = b_i , \qquad \frac{\partial^2 F[\vec{m}]}{\partial m_i \partial m_j} = \frac{\partial b_i}{\partial m_j} = \left(\chi^{-1} \right)_{ij} . \tag{4.52}$$

Using the properties of the inverse Legendre transform, one immediately recovers back $\Omega[\vec{b}]$:

$$-\beta \Omega[\vec{b}] = -\beta \min_{\vec{m}} \left[F[\vec{m}] - \sum_i b_i m_i \right] \tag{4.53}$$

where the minimum over $\vec{m}$ is due to the positivity of the second derivative of $F[\vec{m}]$. By construction, the Legendre transform guarantees the convexity of the functional $F[\vec{m}]$: it cannot develop more local minima that would instead correspond to the metastable states of the system.[4] A promising way to define a non-convex functional and then to study the nature and the number of its metastable states is to compute a high-temperature or small-coupling expansion, according to the formalism which dates back to Plefka [38] and to Georges and Yedidia [27] later. The advantage of dealing with fully-connected models relies on the possibility to truncate the expansion after a finite number of terms, to the first order for a ferromagnetic system, as the Curie-Weiss model, or more generally to the second order for disordered systems. For the sake of convenience, we will define in the following $A^{\beta}[\vec{m}] = -\beta F[\vec{m}]$ and $\lambda_i^{\beta} = \beta b_i$, from which we can write:

$$A^{\beta}[\vec{m}] = \log \sum_{\vec{\sigma}} e^{-\beta[\vec{\sigma}] + \sum_i \lambda_i^{\beta} (\sigma_i - m_i)} . \tag{4.54}$$

The condition for λ_i^{β} implies:

---

[4]An alternative approach to deeper investigate the properties of the free-energy functional consists in performing a perturbative expansion in the inverse of the dimension $1/d$ for large d. This is nothing but a high-temperature expansion of $F[\vec{m}]$. Expanding around this limit—if possible—allows to study the metastable (TAP) states of the system. To be more precise, one should take care of the fact that only local minima of the TAP functional are actually related to real metastable states.

$$\lambda_i^\beta = -\frac{\partial A^\beta[\vec{m}]}{\partial m_i} . \tag{4.55}$$

The first two derivatives with respect to the inverse temperature β are easily computable, given that $m_i = \langle \sigma_i \rangle$, $\forall \beta$:

$$\frac{dA^\beta[\vec{m}]}{d\beta} = \langle -H[\vec{\sigma}] + \sum_i \partial_\beta \lambda_i^\beta (\sigma_i - m_i) \rangle = -\langle H \rangle ,$$

$$\frac{d^2 A^\beta[\vec{m}]}{d\beta^2} = \langle H \left(H[\vec{\sigma}] - \langle H \rangle - \sum_i \partial_\beta \lambda_i^\beta (\sigma_i - m_i) \right) \rangle . \tag{4.56}$$

This brief parenthesis is essentially for better explaining how to correctly develop a free-energy functional in mean-field models, the perceptron model being the key example.

4.4 TAP Equations in the Negative Perceptron

The determination of an effective potential turns out to be a central issue especially in the SAT phase where the energy manifold is flat and unseemly to describe small harmonic fluctuations around the metastable states of the system. In the UNSAT phase, the zero temperature free-energy coincides with the energy. Therefore, energy minima are isolated and their properties can be studied directly [22], recovering the typical glassy features above and at jamming. Conversely, the SAT phase is characterized by a zero energy manifold, meaning that the free energy coincides with the entropy. This phase is identified by a flat energy landscape around the minima and the only possible scenario is a zero Hessian as well. However, the effective potential, i.e. the free-energy as a function of average particle positions in these regions cannot be flat, requiring an alternative and more suitable definition (as shown for a generic rough landscape in Fig. 4.3). In [2] the derivation of an effective potential as a function of local order parameters, namely of the average particle positions and the *generalized forces*, has been proposed. We shall clarify in the following what we mean for generalized forces, which actually stem from the differentiation of the energy with respect to the effective gaps rather than to the particle positions. Our analysis is based on a small-coupling expansion, inspired by the formalism introduced in [27, 38]. We will end up with a non-convex free-energy functional that nevertheless allows us to study metastable states in detail. Thanks to the fully-connected structure of the model, a simplified derivation and a reasonable truncated expansion, taking into account only a finite number of terms, are feasible.

One of the first derivations of the TAP equations for the perceptron dates back to [33]. The author, using the cavity method for a binary model with $\xi_i^\mu = \pm 1$, provided a computation of the number of patterns, which can be stored in an optimal neural

Fig. 4.3 Possible scenario
for the free-energy landscape
in the non-convex
perceptron, in the SAT
phase. Although the energy
landscape is flat, the same
picture does not apply to the
free energy in the SAT phase,
which is very complex at
sufficiently large values of α

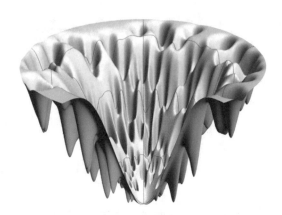

network, which has to be meant as an associative Hopfield-like memory. Our main
purpose in this thesis is nevertheless to deal with critical properties of amorphous
systems at zero temperature, close to the jamming threshold. Recalling Eq. (4.50)
and its Legendre transform in Eq. (4.51), we want to repeat the same procedure in
the perceptron model in order to define an effective potential as a function of mean
particle positions. We start from the definition of the partition function that includes
N additional Lagrange multipliers, u_i, associated with the particle positions x_i:

$$e^{G(\vec{m})} = e^{\sum_{i=1}^{N} m_i u_i} \int d\vec{x} \; e^{-\beta H[\vec{x}] - \sum_{i=1}^{N} x_i u_i} = e^{\sum_{i=1}^{N} m_i u_i + K[\vec{u}]} . \qquad (4.57)$$

This expression is evaluated at the point $\vec{u}$ such that $\vec{m} + \nabla_{\vec{u}} K(\vec{u}) = 0$. $G(\vec{m})$ represents a coarse-grained free-energy after integrating out fast degrees of freedom.
However, to explicitly compute $G(\vec{m})$ we consider worthwhile to write a more generic
expression of the potential, which includes in its definition generalized forces as well.

Given the definition of the gaps in Eq. (4.19), we enforce $h_\mu = h_\mu(x)$ in the
partition function via M auxiliary variables $i\hat{h}_\mu$ conjugated to the gaps. The average
values of the forces and the positions, which the free energy functional actually
depends on, are enforced via the Lagrange multipliers u_i and v_μ.

$$e^{-\Gamma(\vec{m}, \vec{f})} = \int d\vec{x} d\vec{h} d\hat{\vec{h}} \; e^{-\beta H[\vec{h}] + \sum_i (x_i - m_i) u_i + \sum_\mu (i\hat{h}_\mu - f_\mu) v_\mu + \sum_\mu i\hat{h}_\mu (h_\mu(x) - h_\mu)} =$$
$$= e^{J(\vec{u}, \vec{v}) - \vec{m} \cdot \vec{u} - \vec{f} \cdot \vec{v}} , \qquad (4.58)$$

with $\frac{\partial J}{\partial u_i} = \frac{\partial J}{\partial v_\mu} = 0$, $\forall i, \mu$. The functional $\Gamma(\vec{m}, \vec{f})$ then reads:

$$\Gamma(\vec{m}, \vec{f}) = \sum_{i=1}^{N} m_i u_i + \sum_{\mu=1}^{M} f_\mu v_\mu - \log \int d\vec{x} d\vec{h} d\hat{\vec{h}} e^{-\beta H[\vec{h}] + \sum_{i=1}^{N} x_i u_i + \sum_{\mu=1}^{M} i\hat{h}_\mu v_\mu + \sum_{\mu=1}^{M} i\hat{h}_\mu (h_\mu(x) - h_\mu)} .$$
$$(4.59)$$

From the expression above, $G(\vec{m})$ is defined as:

$$G(\vec{m}) = \Gamma(\vec{m}, \vec{f}) \quad \text{evaluated in} \quad \frac{\partial \Gamma(\vec{m}, \vec{f})}{\partial \vec{f}} = 0 . \tag{4.60}$$

The functional (4.59) explicitly depends on generalized forces. Let us see in detail what their physical meaning is. We first define the total force acting on particle i as:

$$F_i = -\frac{dH}{dx_i} = \sum_{\mu=1}^{M} (-h_\mu \theta(-h_\mu)) \frac{dh_\mu}{dx_i} = \sum_{\mu \in C} f_\mu S_{\mu i} , \tag{4.61}$$

where f_μ is the *contact force* and $S_{\mu i} = dh_\mu/dx_i$ is the *dynamical matrix* [23]. The notation $\mu \in C$ stands for those contacts such that $h_\mu < 0$, i.e. the set of unsatisfied constraints. Looking at the derivative of the functional $\Gamma(\vec{m}, \vec{f})$ with respect to the gaps, we recover:

$$\frac{d\Gamma(\vec{m}, \vec{f})}{dh_\mu} = \frac{d}{dh_\mu} \left(\frac{\beta}{2} \sum_{\mu=1}^{M} h_\mu^2 \theta(-h_\mu) \right) + \langle i\hat{h}_\mu \rangle . \tag{4.62}$$

As we are interested in the SAT regime where the gaps are positive-definite, the only surviving term is the ensemble average value $\langle i\hat{h}_\mu \rangle$, which we denote as *generalized force* f_μ. The qualifier *generalized* is thus used to indicate the free energy differentiation with respect to the gap rather than to the actual position.

Similarly to a spin-glass model where the free energy is a function of the overlap value, here the free energy functional depends on the *self-overlap* between two particle configurations, a.k.a the Edwards-Anderson parameter, as well as on the first two moments of the forces, namely:

$$q = \frac{1}{N} \sum_{i=1}^{N} m_i^2 , \qquad r = -\frac{1}{\alpha N} \sum_{\mu=1}^{M} f_\mu^2 , \qquad \tilde{r} = \frac{1}{\alpha N} \sum_{\mu=1}^{M} \langle \hat{h}_\mu^2 \rangle . \tag{4.63}$$

Keeping in mind this notation, Eq. (4.59) can be rewritten as:

$$\Gamma(\vec{m}, \vec{f}) = \sum_i m_i u_i + \sum_\mu f_\mu v_\mu - \log \int d\vec{x} d\vec{h} d\hat{\vec{h}} \; e^{S_\eta(\vec{x}, \vec{h}, \hat{\vec{h}})} , \tag{4.64}$$

$$S_\eta(\vec{x}, \vec{h}, \hat{\vec{h}}) = \sum_i u_i x_i + \sum_\mu i v_\mu \hat{h}_\mu - \lambda \sum_i (x_i^2 - N) - \frac{\beta}{2} \sum_\mu h_\mu^2 \theta(-h_\mu) +$$

$$- i \sum_\mu \hat{h}_\mu (h_\mu - \eta h_\mu(x)) - \frac{b}{2} \sum_\mu (\hat{h}_\mu^2 - \alpha N \tilde{r}) . \tag{4.65}$$

Note that we have introduced two additional parameters compared to Eq. (4.59): λ guarantees the correct normalization on the N-dimensional sphere, while b enforces the second moment of $i\hat{h}_\mu$. The value of the multiplier b is constrained to be $1 - q$ by the saddle point equation $\frac{\partial \Gamma}{\partial \bar{r}} = 0$. What is then the role of η? It denotes a formal parameter to perform a Plefka-like expansion [27, 38] of the free energy, which exactly corresponds to a diagrammatic expansion in the inverse of the dimension $1/N$. We also need to fix the average value of $(i\hat{h}_\mu)^2$ that appears to be very useful to end up with a closed set of equations.

The derivation of the effective potential has to be regarded from a broader perspective. We are, in fact, interested in studying the low-energy phase of the perceptron model at zero-temperature. In the zero-temperature limit two different behaviors can be highlighted, depending on the specific region of the phase space we are interested in. As already discussed in Sect. 4.2.1, in the SAT phase several solutions are possible and the overlap parameter $q < 1$. Conversely, in the UNSAT phase the energy has one single minimum and the overlap parameter is always equal to one. In the following we will focus on the SAT phase in the $T \to 0$ limit. In this regime the free energy corresponds to the configurational entropy of the system as a measure of the number of microstates $\Omega(v)$ with a given volume v. In other terms $S \propto \log \Omega(v)$, where $\Omega(v) = \int d\vec{x}\delta(v - W(\vec{x}))\Theta_{\text{jamm}}$, the Θ function enforcing the excluded volume constraint [19, 20].

The core of our computation properly lies in the definition of an auxiliary *effective Hamiltonian* $\mathcal{H}_{\text{eff}} = i\eta \sum_\mu \hat{h}_\mu h_\mu(\vec{x})$. In a fully-connected system in the large-N limit one can recover mean-field predictions considering only the first two terms of the expansion.[5]

The necessity of introducing an auxiliary Hamiltonian comes from the fact that the original action defined in Eq. (4.18) is zero in the SAT phase. Entering into more details, we need to determine the following quantity:

$$\Gamma(\eta) = \sum_{n=0} \frac{1}{n!} \left.\frac{\partial^n \Gamma}{\partial \eta^n}\right|_{\eta=0} \eta^n , \tag{4.66}$$

where $\Gamma(\eta)$ is the free-energy functional, which for notational simplicity is indicated above as a function of the coupling η only. The role of η is just fictitious: we will formally expand around $\eta = 0$ and then we will set $\eta = 1$ without any loss of generality. The first derivative of the functional with respect to η coincides with the average effective Hamiltonian evaluated in the coarse-grained values:

$$\frac{\partial \Gamma}{\partial \eta} = -\langle H_{\text{eff}} \rangle = -\sum_{i,\mu} \frac{\xi_i^\mu m_i f_\mu}{\sqrt{N}} , \tag{4.67}$$

---

[5] Higher-order terms provide systematic corrections to the mean-field approximation, relevant for short-range interacting models or finite-dimensional ones.

while the second derivative involves both the connected part of the effective Hamiltonian and the partial derivatives of the Lagrange multipliers u_i and v_μ. This gives rise to the *Onsager reaction term* in the Thouless-Anderson-Palmer (TAP) formalism [39]. The second-order contribution then reads:

$$\frac{\partial^2 \Gamma}{\partial \eta^2} = -\left\{ \langle H_{\text{eff}}^2 \rangle - \langle H_{\text{eff}} \rangle^2 + \langle H_{\text{eff}} \left[\sum_i \frac{\partial u_i}{\partial \eta}(x_i - m_i) + \sum_\mu \frac{\partial v_\mu}{\partial \eta}(i\hat{h}_\mu - f_\mu) \right] \rangle \right\} .$$

(4.68)

As far as the second-order term is concerned, in principle one should consider in Eq. (4.68) several mixing terms. We checked that only those with equal indices ($\mu = \nu$, $i = j$) provide a non-vanishing contribution (we refer the interested reader to Appendix A for more details). Other possible terms are:

$$\sum_{i,\mu,\nu} \xi_i^\mu \xi_i^\nu (1 - q) f_\mu f_\nu \qquad \sum_{i,j,\mu} \xi_i^\mu \xi_j^\mu (\alpha\tilde{r} - \alpha r) m_i m_j$$

(4.69)

but they either sum up to zero for the clustering property of the correlation function or they cancel out with an identical term of opposite sign. We are also allowed to neglect for the moment third and higher order terms, which would give a vanishing contribution in the thermodynamic limit [13, 14, 27]. Then the remaining terms to compute are:

$$\frac{\partial u_i}{\partial \eta} = \frac{\partial^2 \Gamma}{\partial \eta \partial m_i} = -i \sum_\mu \frac{\xi_i^\mu \langle \hat{h}_\mu \rangle}{\sqrt{N}} = -\sum_\mu \frac{\xi_i^\mu f_\mu}{\sqrt{N}} ,$$

(4.70)

$$\frac{\partial v_\mu}{\partial \eta} = \frac{\partial^2 \Gamma}{\partial \eta \partial f_\mu} = -\sum_i \frac{\xi_i^\mu \langle x_i \rangle}{\sqrt{N}} = -\sum_i \frac{\xi_i^\mu m_i}{\sqrt{N}} .$$

(4.71)

The information collected so far allows us to rewrite the last terms in Eq. (4.68) as:

$$\langle H \sum_i \frac{\partial u_i}{\partial \eta}(x_i - m_i) \rangle = \alpha N r (1 - q)$$

(4.72)

$$\langle H \sum_\mu \frac{\partial v_\mu}{\partial \eta}(i\hat{h}_\mu - f_\mu) \rangle = \alpha N q (\tilde{r} - r) .$$

(4.73)

Therefore, the full expression for the second derivatives turns out to be:

$$\frac{\partial^2 \Gamma}{\partial \eta^2} = -\{\alpha N(-\tilde{r} + qr) + \alpha N r(1 - q) + \alpha N q(\tilde{r} - r)\} = \alpha N \left[(\tilde{r} - r)(1 - q) \right]$$

(4.74)

and hence the effective potential up to the second order in η:

$$\Gamma(\vec{m}, \vec{f}) = \sum_i \phi(m_i) + \sum_\mu \Phi(f_\mu) + \left.\frac{\partial \Gamma}{\partial \eta}\right|_{\eta=0} \eta + \frac{1}{2}\left.\frac{\partial^2 \Gamma}{\partial \eta^2}\right|_{\eta=0} \eta^2 + \mathcal{O}(\eta^3) =$$

$$\approx -\frac{N}{2}\log(1-q) + \sum_\mu \Phi(f_\mu) - \sum_{i,\mu} \frac{\xi_i^\mu m_i f_\mu}{\sqrt{N}} + \frac{\alpha N}{2}(\tilde{r} - r)(1 - q) .$$

$$(4.75)$$

While for a fully-connected ferromagnetic model the only relevant term is the first moment, since all couplings are $O(1/N)$ and all spins are equivalent [36], in a disordered system, such as the one studied here, both the first and the second moments cannot be neglected at all. More precisely, the second-order term in Eq. (4.75) plays the role of an *Onsager reaction term* [36], describing the fluctuations between $\langle \hat{h}_\mu^2 \rangle$ and $\langle \hat{h}_\mu \rangle^2$.[6] To obtain the last line of Eq. (4.75), we have simply evaluated via a saddle-point approximation the integral over $\vec{x}$, corresponding to the entropy of a non-interacting system constrained on a spherical manifold. It turns out to be proportional to $\log(1 - q)$, similarly to the p-spin spherical model described in Sect. 2.3.1.

$$\phi(m) = \min_u \left[mu - \log \int dx e^{-\lambda(x^2-1)+ux} \right] .$$

$$(4.76)$$

Using the integral representation of the delta-function and neglecting irrelevant prefactors, this computation leads to:

$$\sum_i \phi(m_i) \approx -\frac{N}{2}\log(1-q) .$$

$$(4.77)$$

Moreover, factorizing the terms which depend on the Lagrange multiplier v_μ and $\hat{h}_\mu$, $\hat{h}_\mu^2$ respectively, the functional $\sum_\mu \Phi(f_\mu)$ can be rewritten in a more straightforward way, as follows:

$$\Phi(f) = \min_v \left[fv - \log \int \frac{dh d\hat{h}}{2\pi} e^{\frac{\beta}{2}h^2\theta(-h)-i\hat{h}(h+\sigma)+iv\hat{h}-\frac{b}{2}(\hat{h}^2-\tilde{r})} \right] .$$

$$(4.78)$$

Note that while the integral over $\hat{h}_\mu$ is extended over all values in $(-\infty, \infty)$, the integral over the gaps h_μ can take only positive values in the SAT phase. Since $i\hat{h}_\mu$ is a real variable by definition, namely a physical force, the integration is actually performed in the complex plane and one looks at the values of h_μ and $\hat{h}_\mu$ for which the action is stationary. Reminding that at the saddle-point the parameter b is equal to $(1 - q)$, we finally get:

---

[6]In the Sherrington-Kirkpatrick model, a pivotal example of disordered fully connected system, the analogous expression for the free energy is: $-\beta F[m] = \sum_i s(m_i) + \frac{\beta}{2}\sum_{i\neq j} J_{ij}m_i m_j + \frac{N\beta^2}{4}(1 - q)^2$, where the Onsager term is proportional only to $(1 - q)^2$. Here we find a more complicate situation with a dependence even on the two first moments of the force.

$$\Phi(\vec{f}) = \min_{v}\left[fv - \log H\left(\frac{\sigma - v}{\sqrt{1-q}}\right)\right], \tag{4.79}$$

where we used the notation $H(x) \equiv \frac{1}{2}\mathrm{Erfc}\left(\frac{x}{\sqrt{2}}\right)$ as before. It is worth noticing that by differentiating the expression above with respect to v_μ we immediately get the expression for the forces f_μ. We will get back to this concept later to highlight the importance of leading and sub-leading contributions in terms of f_μ.

Exploiting more Eq. (4.75) we can also derive the stationary equations—in terms of the local quantities m_i and f_μ—that can be solved iteratively:

$$\frac{\partial \Gamma}{\partial m_i} = 0 \quad \Rightarrow \quad m_i\left(\frac{1}{1-q} - \alpha(\tilde{r} - r)\right) = \sum_\mu \frac{\xi_i^\mu f_\mu}{\sqrt{N}}, \tag{4.80}$$

$$\frac{\partial \Gamma}{\partial f_\mu} = \Phi'(f_\mu) - \sum_i \frac{\xi_i^\mu m_i}{\sqrt{N}} + (1-q)f_\mu = 0. \tag{4.81}$$

In an ordinary ferromagnet the solution of these equations is very easy to find since the couplings are known and they do not depend on space indices separately. Conversely, in a spin glass or a generic disordered system the situation is more involved, since the ξ_i^μ's are random variables whose probability distribution is the only available information. However, for an infinite-range model in the thermodynamic limit, it is possible to prove [36] that only a marginal modification is needed, namely to consider an auxiliary system of $N - 1$ and $M - 1$ variables with the i-th and the μ-th ones removed. Using the notation $\sum_i \frac{\xi_i^\mu m_i}{\sqrt{N}} \equiv h_\mu(\vec{m}) + \sigma$ and noticing that $\Phi'(f_\mu) = v_\mu$, we can rewrite Eq. (4.81) as:

$$h_\mu(\vec{m}) = v_\mu - \sigma + (1-q)f_\mu = v_\mu - \sigma - \sqrt{1-q}\,\frac{H'\left(\frac{\sigma - v_\mu}{\sqrt{1-q}}\right)}{H\left(\frac{\sigma - v_\mu}{\sqrt{1-q}}\right)}. \tag{4.82}$$

If the argument of the complementary error function $H(x)$ is much greater than one, i.e. in the jamming limit, the last term can be simplified and the resulting expression turns out to be linear in $(\sigma - v_\mu)/\sqrt{1-q}$ with an opposite sign with respect to the first piece. The two terms cancel out and the jamming limit is exactly identified by the condition $h_\mu \to 0$. The random gaps can be thus expressed as the contribution of the so-called *cavity field* in the spin-glass literature and the Onsager reaction term, the latter giving the correction with respect to the naïve mean-field equation. This argument can be understood taking Eq. (4.81), where the value of v_μ is actually due to m_i in the absence of the μ-th contact. The reaction term, namely $(1-q)f_\mu$, represents instead the influence of the μ-th particle on the others (Fig. 4.4). Therefore, there is a subtle difference between the effective gap $h_\mu(\vec{m})$ and the cavity field $v_\mu - \sigma$, i.e. the field that neighboring particles would feel if a single particle was removed from the system. Hence, the set of values for which $h_\mu < 0$ correspond to the effective contacts

Fig. 4.4 Generalized forces f_μ as a function of $\sigma - v_\mu$ plotted for different values of the overlap q. In the jamming limit, as $q \to 1$, the function approaches the vertical axis (blue), in agreement with the expected divergence of the forces

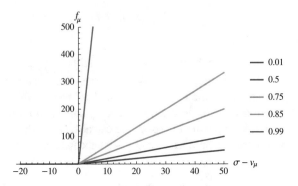

at jamming and, since in the SAT phase the gaps are positive, the only possibility is to have negative values for the cavity field. The method proposed above to derive the TAP formalism for the perceptron is formally similar to that detailed in [7, 46], with the different aim there to focus on optimization and inference problems.

4.4.1   Cavity Method

An alternative way to derive the TAP equations is to consider a graphical representation associating a factor graph [29, 34] to the model defined above. We do not enter into details of the multiplicity of graphical representations and approximations that can be used in this context. The goal of the following pages is to derive the belief propagation (BP) equations [34] for the x_i's and $i\hat{h}_\mu$'s, assuming then that they can be parametrized by Gaussian distributions. The cavity method has been developed in mean-field spin-glass theory as an alternative to the replica method [36]. It is a direct and beneficial tool to study models that are defined on locally tree-like graphs. It is a generalization of the BP approximation, whose underlying idea is that, if a large-size system is increased by single units, its properties do not change too much. We postpone the discussion about the physical meaning of the BP method to Chap. 7 where a quite broad overview of these techniques will be given as an effort to go beyond mean-field approximations. We only highlight here that BP is a message passing procedure useful to write the marginal probability distributions of one variable and then to reduce the problem to the solution of a set of non-linear equations.

In this specific framework, we want to show another method, exactly equivalent to the Plefka-like expansion, which sometimes appears more direct and easier to decode. Given the definition of the effective Hamiltonian in Eq. (4.67), we can write the following message passing equations for this set of messages $\hat{m}_{\mu \to i}(x_i), m_{i \to \mu}(x_i)$ and $\hat{n}_{i \to \mu}(\hat{h}_\mu), n_{\mu \to i}(\hat{h}_\mu)$:

Fig. 4.5 The figure shows the factor graph for the perceptron model, where the circles identify node variables, while the squares represent mutual interactions. Different colors are used to indicate x_i and $i\hat{h}_\mu$

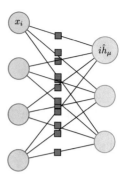

$$\hat{m}_{\mu \to i}(x_i) = \int d\hat{h}_\mu e^{\frac{\xi_i^\mu}{\sqrt{N}} i \hat{h}_\mu x_i} n_{\mu \to i}(\hat{h}_\mu)$$

$$m_{i \to \mu}(x_i) = \prod_{\nu \in \partial i \setminus \mu} \hat{m}_{\nu \to i}(x_i)$$

$$\hat{n}_{i \to \mu}(\hat{h}_\mu) = \int dx_i e^{\frac{\xi_i^\mu}{\sqrt{N}} i \hat{h}_\mu x_i} m_{i \to \mu}(x_i) \qquad (4.83)$$

$$n_{\mu \to i}(\hat{h}_\mu) = \prod_{j \in \partial \mu \setminus i} \hat{n}_{j \to \mu}(\hat{h}_\mu) .$$

The structure of the factor graph for the perceptron is shown in Fig. 4.5, where circles identify variable nodes, i.e. the x_i's the $i\hat{h}_\mu$'s respectively. Black squares instead stand for interactions, also called function nodes. We aim at rewriting the equations for the messages above in a more explicit form. A possible way is to study them in the thermodynamic limit: in this regime we can simplify the equations (*relaxed BP*) assuming a Gaussian distribution to parametrize the messages. Provided that the exponent can be expanded up to order $O(1/N)$, this approximation leads to:

$$\hat{m}_{\mu \to i}(x_i) = e^{\frac{\xi_i^\mu}{\sqrt{N}} a_{\mu \to i} x_i - \frac{1}{2N} (\xi_i^\mu x_i)^2 b_{\mu \to i}} , \qquad (4.84)$$

where

$$a_{\mu \to i} = \int d\hat{h}_\mu n_{\mu \to i}(\hat{h}_\mu) i \hat{h}_\mu ,$$

$$b_{\mu \to i} = \int d\hat{h}_\mu n_{\mu \to i}(\hat{h}_\mu) \hat{h}_\mu^2 - \langle \hat{h}_\mu \rangle^2 . \qquad (4.85)$$

In the Gaussian approximation we can write the message $m_{i \to \mu}(x_i)$ as:

$$m_{i \to \mu}(x_i) = e^{A_{i \to \mu} x_i - \frac{1}{2} B_{i \to \mu} x_i^2} , \qquad (4.86)$$

where

$$A_{i \to \mu} = \frac{1}{\sqrt{N}} \sum_{\nu \in \partial i \backslash \mu} \xi_i^\nu a_{\nu \to i} \; ,$$

$$B_{i \to \mu} = \frac{1}{N} \sum_{\nu \in \partial i \backslash \mu} (\xi_i^\nu)^2 b_{\nu \to i} \; . \tag{4.87}$$

For the other quantities, $\hat{n}_{i \to \mu}(\hat{h}_\mu)$ and $n_{\mu \to i}(\hat{h}_\mu)$, the same treatment can be applied. Then, to recover the TAP formalism, we need to write these equations in terms of single-site and single-pattern marginals, namely:

$$P(x_i) = e^{A_i x_i - \frac{1}{2} B_i x_i^2} \; . \tag{4.88}$$

where we notice that:

$$A_i = A_{i \to \mu} + \frac{1}{\sqrt{N}} \xi_i^\mu a_{\mu \to i} \; ,$$

$$B_i = B_{i \to \mu} + \frac{1}{N} \xi_i^{\mu 2} b_{\mu \to i} \; . \tag{4.89}$$

The key point, which allows to obtain the resulting TAP equations, relies on the assumption that the messages $a_{\mu \to i}$ are roughly equal to A_i up to a small correction, which can be treated perturbatively in the thermodynamic limit. The same conclusions hold for the other set of equations, by simply interchanging the role of x_i with $i\hat{h}_\mu$:

$$\hat{n}_{i \to \mu}(\hat{h}_\mu) = e^{\frac{\xi_i^\mu}{\sqrt{N}} c_{i \to \mu} i \hat{h}_\mu - \frac{1}{2N} (\xi_i^\mu \hat{h}_\mu)^2 d_{i \to \mu}} \; , \tag{4.90}$$

where the first two moments read:

$$c_{i \to \mu} = \int dx_i m_{i \to \mu}(x_i) x_i$$

$$d_{i \to \mu} = \int dx_i m_{i \to \mu}(x_i) x_i^2 - c_{i \to \mu}^2 \; . \tag{4.91}$$

Similarly to Eq. (4.86), we write:

$$n_{\mu \to i}(\hat{h}_\mu) = e^{C_{\mu \to i} i \hat{h}_\mu - \frac{1}{2} D_{\mu \to i} (\hat{h}_\mu)^2} \; , \tag{4.92}$$

where we have again:

$$C_{\mu \to i} = \frac{1}{\sqrt{N}} \sum_{j \in \partial \mu \backslash i} \xi_j^\mu c_{j \to \mu} \; ,$$

$$D_{\mu \to i} = \frac{1}{N} \sum_{j \in \partial \mu \backslash i} \left(\xi_j^\mu \right)^2 d_{j \to \mu} \; . \tag{4.93}$$

A set of equations analogous to (4.89) can be written in this case as well. We thus expand the relative messages in the exponent:

$$n_{\mu \to i}(\hat{h}_\mu) = e^{\left(C_\mu - \frac{1}{\sqrt{N}}\xi_i^\mu c_{i \to \mu}\right)i\hat{h}_\mu - \frac{1}{2}\left(D_\mu - \frac{1}{N}\xi_i^{\mu 2} d_{i \to \mu}\right)\hat{h}_\mu^2} \;, \tag{4.94}$$

which to leading order yields:

$$n_{\mu \to i}(\hat{h}_\mu) \approx n_\mu(\hat{h}_\mu)\left(1 - \frac{1}{\sqrt{N}}\xi_i^\mu c_{i \to \mu}\hat{h}_\mu\right) . \tag{4.95}$$

Then, to determine a closed set of equations, we need to expand the messages $a_{\mu \to i}$ and $b_{\mu \to i}$ in a suitable way, i.e.:

$$\begin{aligned} a_{\mu \to i} &\approx a_\mu - \frac{1}{\sqrt{N}}\delta a_{\mu \to i} = \langle i\hat{h}\rangle_\mu - \frac{1}{\sqrt{N}}\xi_i^\mu c_{i \to \mu}\left(\langle \hat{h}_\mu^2\rangle - \langle \hat{h}_\mu\rangle^2\right) \\ b_{\mu \to i} &\approx b_\mu = \langle \hat{h}_\mu^2\rangle - \langle \hat{h}_\mu\rangle^2 \end{aligned} \tag{4.96}$$

from which we finally obtain:

$$A_i = \frac{1}{\sqrt{N}}\sum_\mu \xi_i^\mu a_\mu - \frac{1}{N}\sum_\mu \xi_i^{\mu 2} c_i\left(\langle \hat{h}_\mu^2\rangle - \langle \hat{h}_\mu\rangle^2\right) , \tag{4.97}$$

$$C_\mu = \frac{1}{\sqrt{N}}\sum_i \xi_i^\mu c_i - \frac{1}{N}\sum_i \xi_i^{\mu 2} d_i \langle i\hat{h}_\mu\rangle . \tag{4.98}$$

Recalling the definition of a_μ, c_i and d_i in Eqs. (4.85) and (4.91) and making use of the cavity method, we can then provide another formulation to confirm the precedent results, as shown in Eqs. (4.80)–(4.81). The two quantities A_i and C_μ are actually related to the previous ones by: $A_i \to \frac{m_i}{1-q}$ and $C_\mu \to \Phi'(f_\mu)$.

It is worth noticing that the TAP equations, largely used in error correcting codes and computer science, can nevertheless get into trouble and do not converge albeit no replica symmetry breaking occurs. Fixed points of the TAP formalism are stationary points of the corresponding TAP free energy and in their derivation from the very beginning they do not include any time indices. Then, if one tries to iterate them naively, that is by putting $t-1$ on all marginals on the right-hand side and t on the left-hand side, their convergence is not assured. This effect is due to a wrong insertion of time indices that actually should include a $t-2$ dependence on the Onsager reaction term. This argument can be immediately understood by Taylor expanding the marginals up to the second order and substituting in the single-site expressions, as Eqs. (4.97) and (4.98). A long explanation of this numerical trouble can be found in [7, 46]. It can nevertheless appear as an artifact of the derivation and be rather involved. An alternative and more intuitive derivation of this time-index dependence dates back to [5], where the author studied dynamically the TAP equations in mean-field spin-glass models, already mentioning the non-Markovian nature of the Onsager reaction term.

4.5 Logarithmic Interaction Near Random Close Packing Density

This Section is devoted to the analysis of the effective potential in the SAT phase close to the jamming threshold. We explain the main steps of the derivation in the perceptron model, but the same analysis can be safely generalized to sphere systems. The topic of this Section is inspired by the argument proposed a few years ago by Brito and Wyart [9, 10] suggesting an analogy between the free energy of a hard-sphere glass and the energy of an athermal network of springs with logarithmic interactions. If the observation time scale of the dynamics is much larger than the collision time but smaller than the relaxation time due to structural rearrangements, one can define a contact network. In this framework two particles (hard spheres) are said to be in contact if they collide within a given time window. Configurations for which touching particles do not overlap are equiprobable [19, 20] and they satisfy the following condition over all contacts $\langle ij \rangle$:

$$\prod_{\langle ij \rangle} \theta \left(\left| \vec{R}_i - \vec{R}_j \right| - \sigma \right) = 1 \,, \tag{4.99}$$

once the network of particles with diameter σ is defined in a metastable state. The partition function can thus be written as:

$$\mathcal{Z} = \int dV \prod_i \int d\vec{R}_i \prod_{\langle ij \rangle} \theta \left(\left| \vec{R}_i - \vec{R}_j \right| - \sigma \right) e^{-\frac{pV}{k_B T}} \,. \tag{4.100}$$

In one dimension the integral (4.100) is exactly solvable by making a variable change from $\vec{R}_i$ to $h_{ij} = \vec{R}_i - \vec{R}_j$, namely:

$$\prod_i d\vec{R}_i \propto \prod_{ij} dh_{ij} \delta \left(\sum_{ij} h_{ij} - (V - V_\infty) \right) \,, \tag{4.101}$$

where V_∞ is the volume evaluated in the infinite pressure limit. Equations (4.100) and (4.101) allow to write:

$$\mathcal{Z} = \prod_{ij} \int_{h_{ij} \geq 0} dh_{ij} e^{-\frac{ph_{ij}}{k_B T}} \,, \tag{4.102}$$

which in turn gives a relation between the pressure and the average gap, inversely proportional one to the other. Generalizing this result in higher dimensions is in principle fairly complex, since the relation between displacements and gaps is no longer linear. However, isostaticity make things easier: thanks to the equality between the number of contacts and the degrees of freedom, this mapping still holds. Near the jamming threshold the linearity is preserved as well and allows to write the following relation:

$$\left(d\vec{R}_i - d\vec{R}_j\right) \cdot \vec{n}_{ij} = dh_{ij} + O(\delta\vec{R}^2) \tag{4.103}$$

where $\vec{n}_{ij}$ represents the unit vector connecting i and j. In higher dimensions the relation for the variable change can be generalized as:

$$\prod_i d\vec{R}_i \propto \prod_{ij} dh_{ij}\delta\left(\sum_{ij} f_{ij}h_{ij} - p\left(V - V_\infty\right)\right), \tag{4.104}$$

where h_{ij} and f_{ij} stand for the gap and the inter-particle force respectively. It can be proven using the virtual work theorem, i.e. $dW = \sum_{ij} f_{ij}dh_{ij} - pdV = 0$ [9, 10], together with the condition that the sum of the forces on all particles must be zero. The generalization of Eq. (4.102) leads to:

$$\mathcal{Z} = \prod_{\langle ij\rangle} \int_{h_{ij}\geq 0} dh_{ij}e^{-\frac{f_{ij}h_{ij}}{k_BT}}, \tag{4.105}$$

which immediately implies a relation between the average gap and the force at jamming:

$$\langle h_{ij}\rangle = \frac{k_BT}{f_{ij}}. \tag{4.106}$$

Therefore, upon approaching the jamming threshold, the forces provide a divergent contribution, which is easy to visualize as the gaps reduce to zero in that regime. We will better explain this point in Sect. 4.7.

By integrating Eq. (4.106) or directly using the definition of the partition function Eq. (4.105), one can check that the thermodynamic potential develops a logarithmic contribution in the average gaps. This insight has long been the subject of analytical and numerical investigations, confirmed in two and three dimensions. Our main goal in the following lines concerns its generalization to every dimension, stressing the importance of universal features lying behind the jamming transition. An exact derivation will be done in the perceptron model although the same conclusions continue to hold for high-dimensional hard spheres.

We refer the reader to Eq. (4.79) reminding that in the SAT phase the functional $\Phi(f)$ of Eq. (4.78) reduces to:

$$\Phi(f) = \min_{v} \left[fv - \log H\left(\frac{\sigma - v}{\sqrt{1-q}} \right) \right].$$

(4.107)

The generalized forces f_μ are the partial derivative of the potential with respect to the Lagrange multiplier v_μ, namely:

$$f_\mu = -\frac{H'\left(\frac{\sigma - v_\mu}{\sqrt{1-q}} \right)}{\sqrt{1-q}\, H\left(\frac{\sigma - v_\mu}{\sqrt{1-q}} \right)},$$

(4.108)

where in the expression for $\Phi(f)$ the dependence on the index μ has been removed taking the minimum over the Lagrange multiplier v. Close to jamming, as $q \to 1$, the argument $\frac{\sigma - v_\mu}{\sqrt{1-q}} \gg 1$ allowing to replace the error function with its asymptotic expansion:

$$H(x) = \frac{1}{2}\mathrm{Erfc}\left(\frac{x}{\sqrt{2}} \right) \approx \frac{e^{-x^2/2}}{\sqrt{2\pi}x}\left(1 + \sum_{n=1}^{\infty}(-1)^n \frac{(2n)!}{n!(\sqrt{2}x)^{2n}} \right),$$

(4.109)

which allows to get a simplified expression for the potential $\Phi(f)$:

$$\Phi(f) = fv - \log H\left(\frac{\sigma - v}{\sqrt{1-q}} \right) \approx fv + \Theta(\sigma - v)\left[\frac{(\sigma - v)^2}{2(1-q)} + \log\left(\frac{\sigma - v}{\sqrt{1-q}} \right) \right].$$

(4.110)

We can simplify the expression for the forces as well, constrained over the value $\sigma - v_\mu$:

$$f_\mu = \Theta(\sigma - v_\mu)\left(\frac{\sigma - v_\mu}{1-q} + \frac{1}{\sigma - v_\mu} \right).$$

(4.111)

Then, we rewrite Eq. (4.81) in the following way:

$$\sigma - v_\mu = -h_\mu(\vec{m}) + (1-q)f_\mu,$$

(4.112)

as a function of the average gap that in shorthand notation reads $h_\mu(\vec{m}) = \frac{\xi^\mu \cdot m}{\sqrt{N}} - \sigma$. Taking into account Eqs. (4.111) and (4.112), we conclude that to leading order the average gap is inversely proportional to the generalized contact force:

$$h_\mu(\vec{m}) = \frac{1-q}{\sigma - v_\mu} \approx \frac{1}{f_\mu}.$$

(4.113)

Note that since the disclosed expansion for the complementary error function $H(x)$ only holds for large values $\frac{\sigma - v_\mu}{\sqrt{1-q}} \gg 1$, this condition can be rewritten in the following form: $\frac{h_\mu(\vec{m})}{\sqrt{1-q}} \ll 1$. It also implies that $\frac{h_\mu(\vec{m})}{1-q} \ll \frac{1}{h_\mu(\vec{m})}$. By eliminating f_μ from the previous equations and expressing all the quantities in terms of $h_\mu(\vec{m})$, the effective potential turns out to be to leading order:

$$G(\vec{m}) \simeq -\frac{N}{2}\log(1-q) + \sum_\mu \Theta(\sigma - v_\mu)\left[\frac{h_\mu(\vec{m})^2}{2(1-q)} + \log\left(-h_\mu(\vec{m}) + \frac{1-q}{h_\mu(\vec{m})}\right) + \right.$$
$$\left. -\frac{1}{2}\log(1-q) + ...\right].$$

(4.114)

We have to take care of the first two terms in square brackets, since one can observe either a quadratic or a logarithmic behavior, depending on the regime to look into. Neglecting irrelevant numerical prefactors and the first entropic term, the leading contribution turns out to be logarithmic in the gap, provided that the expansion holds for small gaps. The final expression thus reads:

$$G(\vec{m}) \simeq -\sum_\mu \Theta(h_\mu(\vec{m})) \log\left(\frac{h_\mu(\vec{m})}{1-q}\right).$$

(4.115)

This result looks interesting for two main reasons: first we obtained an exact derivation within our models of the logarithmic potential first conjectured by Brito and Wyart [10], studying the microscopic cause of rigidity of three dimensional hard-sphere glasses. Second, hard-sphere models provide a very challenging framework to deal with, since the potential is discontinuous and the Hessian of the energy is not well-defined at all. The important turning point is that, instead of describing the model by means of a hard-core potential, which might be complex also for numerical reasons, we can replace it with a smooth logarithmic interaction. We remind that this argument only works in the jamming regime upon approaching the transition line from the SAT phase as in the complementary UNSAT phase a harmonic interaction would otherwise dominate the free energy.

4.6 Third Order Corrections to the Effective Potential

In the previous Section we explained the derivation of the TAP free energy taking only the first two terms of the expansion into account according to a mean-field-like picture. One could be interested in defining a modified version of the perceptron—for instance a diluted model with finite-connectivity patterns ξ^μ—or even a finite dimensional system not exactly at jamming. In both cases higher-order corrections would be relevant to providing a finite contribution in a perturbative expansion in the inverse of the dimension. In principle, one should consider all the contributions coming from loopy structures by summing over triplets, quadruplets and generic combinations of links. Our computation aims at achieving a better understanding of the deviation of the coarse-grained potential from its critical trend. Therefore, the main expression to evaluate is [27, 35]:

$$\frac{\partial^3 \Gamma}{\partial \eta^3} = \langle H_{\text{eff}} \rangle \frac{\partial \langle H_{\text{eff}} \rangle}{\partial \eta} + \langle H_{\text{eff}} \Upsilon_2 \rangle + \langle H_{\text{eff}} (H_{\text{eff}} - \langle H_{\text{eff}} \rangle + \Upsilon_1)^2 \rangle \qquad (4.116)$$

where Υ_n reads:

$$\Upsilon_n = \sum_i \frac{\partial}{\partial y_i} \left(\frac{\partial^n \Gamma}{\partial \eta^n} \right) (s_i - y_i) . \qquad (4.117)$$

For simplicity, we indicated both derivatives, with respect to m_i and to f_μ, as $(s_i - y_i)\frac{\partial}{\partial y_i}$. Then, the computation of the third-order corrections implies:

$$\frac{\partial^3 \Gamma}{\partial \eta^3} = -\langle H_{\text{eff}} \rangle \frac{\partial^2 \Gamma}{\partial \eta^2} + \langle H_{\text{eff}} \Upsilon_2 \rangle + \langle H_{\text{eff}} (H_{\text{eff}} - \langle H_{\text{eff}} \rangle + \Upsilon_1)^2 \rangle =$$

$$= \langle H_{\text{eff}} \rangle \left[\langle H_{\text{eff}}^2 \rangle - \langle H_{\text{eff}} \rangle^2 - \langle H_{\text{eff}} \sum_i (s_i - y_i) \frac{\partial \langle H_{\text{eff}} \rangle}{\partial y_i} \rangle \right] +$$

$$+ \langle H_{\text{eff}} \sum_i (s_i - y_i) \frac{\partial}{\partial y_i} \frac{\partial^2 \Gamma}{\partial \eta^2} \rangle + \left\langle H_{\text{eff}} \left(H_{\text{eff}} - \langle H_{\text{eff}} \rangle - \sum_i \frac{\partial \langle H_{\text{eff}} \rangle}{\partial y_i} (s_i - y_i) \right)^2 \right\rangle . \qquad (4.118)$$

The expression above can be further simplified, leading to:

$$\frac{\partial^3 \Gamma}{\partial \eta^3} = \langle H_{\text{eff}}^3 \rangle + \langle H_{\text{eff}} \rangle \langle H_{\text{eff}}^2 \rangle - 2\langle H_{\text{eff}} \rangle^3 - \langle H_{\text{eff}} \rangle \alpha N r (1 - q) - \langle H_{\text{eff}} \rangle \alpha N q (\tilde{r} - r) +$$

$$+ \left\langle H_{\text{eff}} \left(-\sum_{i,\mu} \frac{\delta x_i}{\sqrt{N}} \xi_i^\mu f_\mu \right)^2 \right\rangle + \left\langle H_{\text{eff}} \left(-\sum_{i,\mu} \frac{\delta f_\mu}{\sqrt{N}} \xi_i^\mu m_i \right)^2 \right\rangle +$$

$$- 2\langle H_{\text{eff}}^2 \left(\sum_i \delta x_i \sum_\mu \frac{\xi_i^\mu f_\mu}{\sqrt{N}} + \sum_\mu \delta f_\mu \sum_i \frac{\xi_i^\mu m_i}{\sqrt{N}} \right) \rangle . \qquad (4.119)$$

We denote as $\delta x_i = (x_i - m_i)$ and $\delta f_\mu = (i\hat{h}_\mu - f_\mu)$ the relative deviations of the particle positions and the contact forces from their own mean value. The first terms in Eq. (4.119) reminds an analogous expression that can be written for the Sherrington-Kirkpatrick model using a TAP approach [36]. The other terms are instead due to the variation of the additional parameters on which the perceptron depends.

Close to jamming, signaled by the condition $q \to 1$, most of these terms can be re-expressed in a more straightforward way and simplified with each other. The fourth and the fifth terms cancel out with the next two terms having opposite sign, their squared moments being $(1 - q)r$ and $(\tilde{r} - r)q$ respectively. Hence, we have to focus only on the first three terms and the very last one. The last term is not relevant in the expansion, due to the fact that in the jamming limit the actual values of the position and the force tend to their coarse-grained values, i.e. $x_i \to m_i$ and $i\hat{h}_\mu \to f_\mu$. The most interesting contribution comes from the first three terms. Thanks to the fact that the jamming transition is characterized by very weak correlations, a direct connection between the jamming transition and a mean-field-like scenario can be pointed out. Using this reasoning one can show that the first three terms in the jamming limit cancel out providing a total vanishing contribution in the TAP expansion. This result is in remarkable agreement with the argument asserting that the jamming transition is well described in terms of binary interactions only [9, 26].

We also expect that fourth-order corrections to the effective potential do not change the disclosed behavior in the critical jamming region: the underlying reason is again related to isostaticity.

It is worth noticing that in glassy systems a perturbative diagrammatic expansion of the correlation functions can be established if the particle cages are sufficiently small [37], namely in the high-pressure limit. The hypernetted chain HNC approximation [16, 28] can be considered the simplest method to treat correlation functions in the liquid phase. The main objective is to find a solution for the correlator $g_{ab}(x, y)$, representing the probability to find a particle of type b in y conditioned to the fact that a particle of type a is in x. However, the main difficulty stems from the fact that the HNC theory fails for small cage radius. In fact, it takes into account only the less divergent diagrams, which are not the only relevant contributions deeply into the glass phase. To correctly treat the jamming transition, one should take care also of the maximally divergent diagrams which have a completely connected structure. Alternative approaches must be used. In particular, in [37] the authors propose a method which allows to write the correlation functions of the glass as the correlation functions of the effective liquid. In that case, the contributions due to three-point correlators can be factorized and rewritten as a function of two-point correlators only. Our result, based on the determination of a well-defined potential exclusively in terms of the first two moments in the jamming limit, seems to be directly correlated to this issue.

4.7 Leading and Subleading Contributions to the Forces Near Jamming

Despite the fact that inter-particle forces in glassy materials are generally hard to determine experimentally, the force distribution can be exactly reconstructed analytically, at least in the jamming limit. In Sect. 4.5 we highlighted the connection between the effective forces and the gaps in the jamming regime.

The main difficulty in determining the effective interactions in amorphous systems stems from the impossibility of identifying a simple relationship between forces and gaps as soon as one attempts to extend the formalism beyond jamming. Indeed, upon decreasing the density, there is no reason to believe that the effective forces should remain binary. In particular, two different kinds of corrections to the generalized forces might emerge, one due to finite-size corrections and a second one related to computing all observables in the system at a finite distance from jamming. Let us focus on finite-size corrections first. As in the jamming limit Eq. (4.119) reduces to zero, its derivatives with respect to f_μ turn out to be trivially zero at any level, confirming the starting hypothesis that Eq. (4.81) still holds. This means that nothing is going to change in the generic force expression.

The next step is to investigate how the mutual relation between the forces and the gaps is modified upon increasing the distance from jamming, both in a mean-field scenario and in a finite system. We intend to refine our estimate compared to the previous results taking into account also sub-leading terms in the asymptotic expansion of the potential. Inserting the next subleading terms in the expansion of the potential $\Phi(f)$, we get:

$$\Phi(f) \approx \min_v \left\{ fv + \theta(\sigma - v) \left[\frac{(\sigma - v)^2}{2(1 - q)} + \log\left(\frac{\sigma - v}{\sqrt{1 - q}}\right) - \log\left(1 - \frac{1}{[(\sigma - v)/\sqrt{1 - q})]^2}\right)\right]\right\}.$$
(4.120)

Differentiating with respect to v as before, we get a better approximation for the generalized forces:

$$f_\mu = \frac{\sigma - v_\mu}{1 - q} + \frac{1}{\sigma - v_\mu} - \frac{2(1 - q)}{(\sigma - v_\mu)^3 \left[1 - \frac{(1-q)}{(\sigma - v_\mu)^2}\right]}.$$
(4.121)

Two different regimes can be highlighted depending on the order of magnitude of $\frac{\sigma - v_\mu}{\sqrt{1-q}}$. If $\frac{\sigma - v_\mu}{\sqrt{1-q}} \ll 1$, the contribution due to the logarithmic term in Eq. (4.107) vanishes, as correctly expected by the definition of the error function. Going ahead and assuming that q is close to its maximum value, we can expand the last term as a sum of odd powers of $\sigma - v_\mu$, yielding:

$$f_\mu \approx \left(\frac{\sigma - v_\mu}{1 - q}\right)\left[1 + \frac{1 - q}{(\sigma - v_\mu)^2} - \frac{2(1 - q)^2}{(\sigma - v_\mu)^4} - \frac{2(1 - q)^3}{(\sigma - v_\mu)^6} + \dots\right]$$

$$\approx \frac{1}{h_\mu(\vec{m})}\left[1 + \frac{h_\mu(\vec{m})^2}{1 - q} - \frac{2h_\mu(\vec{m})^4}{(1 - q)^2} + \dots\right]$$
(4.122)

$$f_\mu \approx \frac{1}{h_\mu(\vec{m})} \mathcal{G}\left(\frac{h_\mu(\vec{m})}{\sqrt{1 - q}}\right).$$

The intermediate expression in Eq. (4.122) is justified by the fact that at the leading order the term $\frac{\sigma - v}{1 - q}$ coincides with the inverse gap.[7] The last line of Eq. (4.122) can be easily understood by looking at the connected part of the average gap, namely:

$$\langle h_\mu^2 \rangle_c = \frac{1}{N} \sum_{ij} \xi_i^\mu \xi_j^\mu \left(\langle x_i x_j\rangle - m_i m_j\right) = 1 - q.$$
(4.123)

According to this relation, Eq. (4.122) can be written in a more compact way:

---

[7] The subsequent terms including odd powers of $\frac{1}{\sigma - v}$ seem to encode the effect of a rescaled inverse pressure, which typically vanishes in the SAT region.

$$f_\mu \approx \frac{1}{\langle h_\mu \rangle} \mathcal{G}\left(\frac{\langle h_\mu \rangle}{\langle h_\mu^2 \rangle_c^{1/2}}, \alpha \right), \tag{4.124}$$

which determines two different behaviors: as $\alpha \to \alpha_J$ the scaling function $\mathcal{G} \to 1$, confirming that the only relevant scale is the inter-particle gap, whereas if $\alpha < \alpha_J$ a full expression for $\mathcal{G}$ is needed. The crossover regime determining where the logarithmic approximation fails is highlighted by the condition: $h_\mu \sim \sqrt{1-q}$. A possible way to better visualize this reasoning is to consider directly the expression for the generalized forces f_μ in Eq. (4.108). Then, making use of the connection existing between forces and average gaps at jamming, i.e. Eq. (4.82), we can write:

$$\sqrt{1-q}\,f_\mu = -\frac{H'\left(\frac{-h_\mu}{\sqrt{1-q}} + \sqrt{1-q}\,f_\mu\right)}{H\left(\frac{-h_\mu}{\sqrt{1-q}} + \sqrt{1-q}\,f_\mu\right)}. \tag{4.125}$$

Inverting this function with respect to f_μ we can immediately get the typical trend shown in Fig. 4.6, divergent as the gaps shrink to zero and finite otherwise. Equation (4.122) and consequently Eq. (4.125) actually suggest a deep analogy between the free energy of a hard-sphere glass and that of an athermal network of logarithmic springs [9, 10], if looking at the dynamics on a time interval much greater than the collisional time but smaller than the structural relaxation time. This leads to the determination of a contact network and, thanks to the fact that all configurations are equiprobable at jamming, a one-to-one mapping between the particle displacements and the gaps can be established. In other words, the total number of contacts equals the number of degrees of freedom, satisfying the isostaticity condition. In this case, a simple relationship between forces and gaps can be established as well. One could nevertheless wonder why this relation should be valid in dimensions higher than one, where the mapping is no more linear. The answer lies in the underlying isostaticity condition which characterizes the jamming transition. However, upon increasing the distance from the jamming line this condition is no longer valid and the forces are not only functions of h_μ but of a complex combination of random parameters.

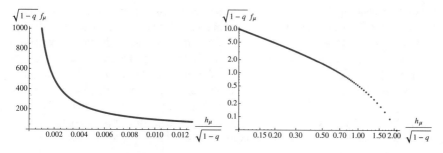

Fig. 4.6 Scaling function showing generalized forces versus gaps, in linear and log-log scales, well interpolated by a linear function in the small-gap regime

In the literature no analytical predictions about the typical scaling of the forces taking into account also sub-leading terms thus far are available. Several numerical simulations have been carried out attempting to explain the observed behavior [26, 43]. In particular, in [42, 43] the authors showed that deviation of the force from its leading behavior can be estimated numerically in molecular dynamics and the next sub-leading contribution should be of order of the number of effective contacts $\delta z = z - z_c$.

4.7.1 Scalings and Crossover Regimes

The argument mentioned a few lines above is also correlated to the interesting problem of matching between the SAT and the UNSAT phases. We discussed in Sect. 4.7 the leading behavior of the forces near jamming and the emergence of a smooth logarithmic interaction in the perceptron model, exactly derivable for high-dimensional systems. However, as $h_\mu \sim \sqrt{1-q}$ the transition towards a logarithmic regime is progressively smeared out. Now we want to address the problem of identifying the appropriate scaling functions in order to describe the different phases of the perceptron phase diagram. We shall exploit the main predictions of the full RSB solution largely explained in [23] to figure out the crossover temperature-dependent behavior between these two phases.

At low temperature in the UNSAT phase the overlap has a simple dependence on temperature given by:

$$1 - q = \chi T + O(T^2) , \qquad (4.126)$$

where the parameter χ is determined by the condition:

$$\left(1 + \frac{1}{\chi}\right) = \sqrt{\frac{\alpha}{\alpha_J(\sigma)}} . \qquad (4.127)$$

The critical value $\alpha_J(\sigma)$ on the jamming line is essentially related to the inverse of an appropriate error function. However, the zero-temperature limit should be carefully performed, sending T and $1 - q = \chi T$ to zero at the same time. The jamming limit is characterized by the conditions: $\chi \to \infty$ and $q \to 1$, which determine two different scaling solutions depending on the values of q and q^*, the Edwards-Anderson parameter and the matching point respectively. The matching point corresponds to the condition $\chi P(1, 0)\sqrt{1 - q^*} \sim 1$, where the probability distribution $P(q, h)$ is evaluated in $q = 1$ and $h = 0$ and verifies the Parisi equation [36]. When $q \gg q^*$ we recover the ordinary UNSAT phase, while as $q \ll q^*$ the jamming solution takes place. We know that in the UNSAT phase the pressure is proportional to the first moment of the gap $[h]$, which in turn satisfies the following relation $[h] \equiv 1/N \sum_{\mu=1}^{M} h_\mu \theta(-h_\mu) \propto 1/\chi^2$. In the end we have [23]:

$$(1 - q^*) \sim \chi^{\frac{k}{1-k}} \qquad (4.128)$$

with a critical exponent $k \approx 1.41$. Our argument is based on the fact that close to jamming the fullRSB equations show a scaling regime. In particular, we focus on the regime in which the Edwards-Anderson parameter is roughly equal to the cut-off value, such that $q \sim q^*$, and we analyze the typical temperature behavior. Let us suppose that the temperature is raised by a finite amount, which enables us to perturb the solution as:

$$(1 - q^*) \sim \chi^{\frac{k}{1-k}} \sim \chi T \tag{4.129}$$

and to get:

$$T^* \sim \chi^{\frac{2k-1}{1-k}} . \tag{4.130}$$

Given that in the soft-sphere regime (UNSAT) the pressure scales as $p \sim 1/\chi^2$ [22, 23], we also obtain:

$$T^* \sim p^{\frac{2k-1}{2k-2}} . \tag{4.131}$$

For $T \sim T^*$ the UNSAT phase and the jamming solution can be no longer distinguished. Note that the relation above connecting temperature and pressure exactly coincides with that derived in [17] based on an Effective Medium Theory (EMT) argument.

On the other hand, in the SAT phase $(1 - q) \sim \epsilon^k$, where ϵ stands for the linear distance from the jamming line. These two relations together lead to the condition:

$$T^* \sim \epsilon^{2k-1} . \tag{4.132}$$

Under these assumptions we should be able to define a scaling function:

$$(1 - q) \sim \epsilon^k \mathcal{F}\left(T \epsilon^{1-2k}\right) , \tag{4.133}$$

which guarantees the correct trend in each regime, either when its argument diverges or goes to zero. According to this simple argument, three different regimes can be highlighted: a HS/SAT regime, characterized by a zero energy manifold and studied here by means of the TAP formalism, a SS/UNSAT regime, whose low-energy vibrational properties have been largely analyzed in [22], and an *anharmonic* regime signaled by the crossover temperature T^* that can also be related to the linear distance ϵ from jamming. Below this threshold the system actually consists of an assembly of soft harmonic particles, whereas above its vibrational properties turn out to be indistinguishable from those of a hard-sphere system.

4.8 Spectrum of Small Harmonic Fluctuations

We largely discussed the implications of a coarse-grained free energy in terms of generalized forces, which are typically not so difficult to determine in a simulation. Another reasonable question concerns the determination of the Hessian of this effective potential. Indeed, from the differentiation of the free-energy functional one can define a dynamical matrix, already introduced in Chap. 3, useful to describe linear responses of particle displacements to any external force. Given the distribution of the eigenvalues of the dynamical matrix, one can also have access to the spectrum of small harmonic fluctuations around each metastable state. We recall the expression of the free-energy functional, depending on both m_i and f_μ:

$$\Gamma(\vec{m}, \vec{f}) \approx -\frac{N}{2} \log(1 - q) + \sum_\mu \Phi(f_\mu) - \sum_{i,\mu} \frac{\xi_i^\mu m_i f_\mu}{\sqrt{N}} + \frac{\alpha N}{2}(\tilde{r} - r)(1 - q)$$

(4.134)

from which the generic expression of the Hessian matrix reads:

$$\frac{d^2\Gamma(\vec{m}, \vec{f})}{dm_i dm_j} = \frac{\partial^2\Gamma}{\partial m_i \partial m_j} + \sum_{\mu=1}^M \frac{\partial^2\Gamma}{\partial f_\mu \partial m_j}\frac{\partial f_\mu}{\partial m_i} + \sum_{\mu=1}^M \frac{\partial^2\Gamma}{\partial f_\mu \partial m_i}\frac{\partial f_\mu}{\partial m_j} + \sum_{\mu,\nu=1}^M \frac{\partial^2\Gamma}{\partial f_\mu \partial f_\nu}\frac{\partial f_\mu}{\partial m_i}\frac{\partial f_\nu}{\partial m_j}.$$

(4.135)

We also recall the following stationary condition:

$$\frac{\partial\Gamma(\vec{m}, \vec{f})}{\partial f_\mu} = \Phi'(f_\mu) - \sum_i \frac{\xi_i^\mu m_i}{\sqrt{N}} + (1 - q)f_\mu = 0$$

(4.136)

that, differentiated with respect to m_i, gives:

$$\left[\Phi''(f_\mu) + (1 - q)\right]\frac{\partial f_\mu}{\partial m_i} - \frac{\xi_i^\mu}{\sqrt{N}} = 0 .$$

(4.137)

From Eq. (4.136) we can also write:

$$\frac{\partial^2\Gamma(\vec{m}, \vec{f})}{\partial f_\mu^2}\frac{\partial f_\mu}{\partial m_i} + \frac{\partial^2\Gamma(\vec{m}, \vec{f})}{\partial f_\mu \partial m_i} = 0 ,$$

(4.138)

which will be very helpful to cancel out mixed and rather complicated terms in the resulting expression of the Hessian. We can thus consider the mixed partial derivative:

$$\frac{\partial^2\Gamma(\vec{m}, \vec{f})}{\partial f_\mu \partial m_j} = -\frac{\xi_j^\mu}{\sqrt{N}} ,$$

(4.139)

which together with:

$$\frac{\partial^2\Gamma(\vec{m}, \vec{f})}{\partial f_\mu f_\nu} = \delta_{\mu\nu}\left[\Phi''(f_\mu) + (1 - q)\right] ,$$

(4.140)

$$\frac{\partial^2\Gamma(\vec{m}, \vec{f})}{\partial m_i \partial m_j} = \delta_{ij}\left[\frac{1}{1 - q} - \alpha(\tilde{r} - r)\right] + \frac{2m_i m_j}{N(1 - q)^2}$$

(4.141)

provide all necessary relations to obtain a simplified expression for the Hessian matrix. In the following we assume to neglect the projector term $\frac{2m_i m_j}{N(1-q)^2}$ and all other $O(1/N)$ terms: if included in the computation, the projector splits off a single isolated eigenvalue from the continuous band [3]. As a consequence, the Hessian turns out to be:

$$
\begin{aligned}
\mathcal{M}_{ij} \equiv \frac{\partial^2 G(\vec{m})}{\partial m_i \partial m_j} &= \delta_{ij} \left[\frac{1}{1-q} - \alpha(\tilde{r} - r) \right] - \frac{2}{N} \sum_{\mu=1}^{M} \left[\Phi''(f_\mu) + (1-q) \right]^{-1} \xi_i^\mu \xi_j^\mu + \\
&\quad + \frac{1}{N} \sum_{\mu=1}^{M} \left[\Phi''(f_\mu) + (1-q) \right]^{-1} \xi_i^\mu \xi_j^\mu = \\
&= \delta_{ij} \left[\frac{1}{1-q} - \alpha(\tilde{r} - r) \right] - \sum_{\mu=1}^{M} \frac{\xi_i^\mu \xi_j^\mu}{N} \frac{1}{\Phi''(f_\mu) + (1-q)} ,
\end{aligned}
$$

(4.142)

consisting of a diagonal term $\zeta = \frac{1}{1-q} - \alpha(\tilde{r} - r)$, which only denotes an extra shift in the spectrum, and an off-diagonal term multiplying the covariance matrix $\xi_i^\mu \xi_j^\mu$. The latter represents the stiffness parameter, i.e.:

$$
k_\mu = -\frac{\partial f_\mu}{\partial h_\mu} = -\frac{1}{\Phi''(f_\mu) + (1-q)}.
$$

(4.143)

Close to jamming $q \to 1$ and this requires the Hessian to be rescaled by $(1-q)$ in order to have finite eigenvalues. This rescaling is justified by the definition of the effective potential, which corresponds indeed to the vibrational entropy of the system. It actually describes the volume of phase space around a given metastable state, the same reason why it can be always expressed in terms of mean particle positions within such states.

4.8.1 Spectrum in the UNSAT Phase

The situation is quite different compared to the complementary UNSAT phase where one can consider the following Hamiltonian [22]:

$$
\mathcal{H}_\zeta[\vec{x}] = \mathcal{H}[\vec{x}] - \frac{N}{2} \zeta(\vec{x}^2 - 1) ,
$$

(4.144)

which, minimized with respect to the particle displacement, gives:

$$
\frac{\partial \mathcal{H}_\zeta}{\partial x_i} = \sum_{\mu=1}^{M} \xi_i^\mu h_\mu \theta(-h_\mu) - N\zeta x_i = 0 .
$$

(4.145)

Then, the Hessian matrix, normalized by N, will be:

$$\mathcal{M}_{ij}^{\text{UNSAT}} = \frac{1}{N} \frac{\partial^2 \mathcal{H}_{\zeta}}{\partial x_i \partial x_j} \approx \frac{1}{N} \sum_{\mu=1}^{M} \xi_i^{\mu} \xi_j^{\mu} \theta(-h_{\mu}) - \zeta \delta_{ij} . \tag{4.146}$$

The UNSAT phase is characterized by a non-vanishing fraction of negative gaps, which contribute to the θ function. In principle, we should take into account the correlations between the covariance matrix and the Heaviside theta. However, in the thermodynamic limit these correlations can be neglected, since each h_{μ}, which the θ function depends on, are a sum of a large number of ξ_i^{μ}. Therefore the Hessian has the typical form of a random matrix from a modified Wishart ensemble with N [1] random contributions[8] and an additional scalar term ζ:

$$\mathcal{M}_{ij}^{\text{UNSAT}} \approx \frac{1}{N} \sum_{\mu=1}^{N[1]} \xi_i^{\mu} \xi_j^{\mu} - \zeta \delta_{ij} = [1] \, W_{ij} - \zeta \delta_{ij} . \tag{4.147}$$

As a consequence, the eigenvalue distribution follows a Marchenko–Pastur law [30], from which two different behaviors can be highlighted, depending on whether the system is hyperstatic or hypostatic:

$$\rho(\lambda) = \begin{cases} (1 - [1]) \delta(\lambda + \zeta) + \nu(\lambda) & [1] < 1 \\[2mm] \frac{1}{2\pi} \frac{\sqrt{(\lambda - \lambda_-)(\lambda_+ - \lambda)}}{\lambda + \zeta} \mathbf{1}_{\lambda_-, \lambda_+}(\lambda) & [1] > 1 . \end{cases} \tag{4.148}$$

A minimum of the Hamiltonian is identified by the condition that all the eigenvalues are positive or zero. According to Eq. (4.148) we can focus on two main scenarios: (i) [1] < 1 implies $\zeta \leq 0$, which is possible if $\sigma > 0$; (ii) if [1] ≥ 1, σ can take only negative values. In the second case the lower edge of the spectrum λ_- can be positive or zero, which in turn leads to the following condition:

$$(\sqrt{[1]} - 1)^2 \geq \zeta = [h^2] + \sigma[h] . \tag{4.149}$$

The edges of the spectrum are thus identified by $\lambda_{\pm} = (\sqrt{[\lambda]} \pm 1)^2 - \zeta$. Generally, the eigenvectors of a Wishart matrix are delocalized on the sphere and this is exactly the same scenario occurring to the Hessian in the perceptron model [22].

The spectral density in the UNSAT phase is shown in Fig. 4.7. As explained before, at zero temperature a transition from a RS phase for $\sigma > 0$, with a gapped spectrum and a unique minimum in the energy landscape, to a full RSB phase for $\sigma < 0$, with multiple minima and a marginally stable spectrum, occurs. Let us see in more details what this picture means. One can expand the Hamiltonian around a minimum in the harmonic approximation, assuming that $x = x^0 + \delta x$:

---

[8]The symbol [1] denotes indeed the fraction of contacts.

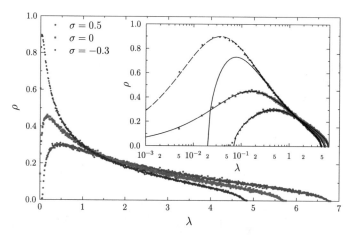

Fig. 4.7 Spectrum of the Hessian in the UNSAT region for $N = 1600$, $\alpha = 4$ and $\sigma = 0.5, 0, -0.3$, in linear (main panel) and semilog (inset) scales. The Marchenko–Pastur law with RS parameters (full lines) perfectly reproduces the data for $\sigma \geq 0$, while deviations are observed for $\sigma < 0$. Reprinted from [22]

$$H[\delta x] = \frac{N}{2}\sum_{ij}\delta x_i M_{ij}\delta x_j - \frac{N}{2}\delta\zeta\sum_i\delta x_i^2 = \frac{N}{2}\sum_j\lambda_j\delta x_j^2 - \frac{N}{2}\delta\zeta\sum_j\delta x_j^2$$

$$(4.150)$$

where $\delta\zeta$ denotes an infinitesimal change in the shift parameter ζ due to a corresponding temperature change. Then the susceptibility can be written as:

$$\chi = \beta(1 - q) = \sum_j\langle\delta\xi_j^2\rangle = \sum_j\frac{T}{N\lambda_j} = \int d\lambda\frac{\rho(\lambda)}{\lambda}.$$

$$(4.151)$$

For positive σ, i.e. in the RS phase, we have $\rho(\lambda) \sim \sqrt{\lambda - \lambda_-}$ with $\lambda_- \sim \sigma^2$ and the lower edge of the spectrum is always positive. Conversely, for negative σ, the model is characterized by a zero lower edge, which results in a gapless spectrum.

4.8.2 Spectrum in the SAT Phase

After discussing physical interpretations of the UNSAT phase, which can be extensively recovered in [22], we want now to focus on the analysis of the SAT phase, where, in principle, the underlying scenario is more difficult to investigate. The SAT phase is exactly the focal point of this part of the thesis. Let us start with the definition of the average resolvent $R(z)$ [32] associated with the Hessian. Then we introduce replicas and, after taking the thermodynamic limit, the analytical continuation to $n \rightarrow 0$ is considered.

$$R(z) = \frac{1}{N}\langle Tr\,(z\mathbf{I} - \mathbf{M})^{-1}\rangle\,, \tag{4.152}$$

where $\mathbf{I}$ is the identity matrix, while $z = \lambda - i\epsilon$ is a factor depending on the regularized infinitesimal parameter $\epsilon > 0$, to be sent to zero at the end. The density $\rho(\lambda)$ can be computed from a limit procedure on the resolvent:

$$\rho^{(N)}(\lambda) = \frac{1}{\pi}\lim_{\epsilon \to 0^+}\Im(R^{(N)}(z))\,. \tag{4.153}$$

We start the detailed computation by means of the replica method, where $a = 1, \dots n$.

$$
\begin{aligned}
R(z) = \lim_{n\to 0}\frac{1}{N}\left(-\frac{2}{n}\right)\frac{\partial}{\partial z}\mathbb{E}_{\vec{\xi}}\int_{-\infty}^{+\infty}\prod_{a=1}^{n}\prod_{i=1}^{N}\left(\frac{d\phi_i^{(a)}}{\sqrt{2\pi/i}}\right)\cdot \\
\exp\left[-\frac{i}{2}\sum_{a=1}^{n}\sum_{i,j=1}^{N}\phi_i^{(a)}\left(z\delta_{ij} - \frac{1}{N}\sum_{\mu=1}^{M}\xi_i^{\mu}\xi_j^{\mu}k_{\mu}\right)\phi_j^{(a)}\right].
\end{aligned}
\tag{4.154}
$$

We assume that the μ-dependent factors k_μ are uncorrelated being their probability distribution perfectly factorized over μ. It can seem a quite rough assumption, which is nevertheless in good agreement with the main predictions by Wyart and coworkers based on marginal stability. By the definition $s_a^\mu \equiv \frac{1}{\sqrt{N}}\sum_{i=1}^{N}\xi_i^\mu\phi_i^{(a)}\sqrt{k_\mu}$, the resolvent can be rewritten as follows:

$$
\begin{aligned}
R(z) = \lim_{n\to 0}\frac{1}{N}\left(-\frac{2}{n}\right)\frac{\partial}{\partial z}\mathbb{E}_{\vec{\xi}}\int_{-\infty}^{+\infty}\prod_{a=1}^{n}\prod_{i=1}^{N}\frac{d\phi_i^{(a)}}{\sqrt{2\pi/i}} \\
\int_{-\infty}^{+\infty}\prod_{a=1}^{n}\prod_{\mu=1}^{M}\frac{ds_\mu^{(a)}d\hat{s}_\mu^{(a)}}{2\pi}\exp\left[-\frac{iz}{2}\sum_{i=1}^{N}\sum_{a=1}^{n}(\phi_i^{(a)})^2 + \frac{i}{2}\sum_{a=1}^{n}\sum_{\mu=1}^{M}(s_\mu^{(a)})^2+\right. \\
\left. + i\sum_{a=1}^{n}\sum_{\mu=1}^{M}s_\mu^{(a)}\hat{s}_\mu^{(a)} - \frac{i}{\sqrt{N}}\sum_{a=1}^{n}\sum_{\mu=1}^{M}\sum_{i=1}^{N}\xi_i^{\mu}\phi_i^{(a)}\hat{s}_\mu^{(a)}\sqrt{k_\mu}\right].
\end{aligned}
\tag{4.155}
$$

The average over the Gaussian distributed disorder leads to:

$$\mathbb{E}_{\vec{\xi}}\exp\left[-\frac{i}{\sqrt{N}}\sum_{a=1}^{n}\sum_{\mu=1}^{M}\sum_{i=1}^{N}\frac{\xi_i^\mu}{\sqrt{N}}\phi_i^{(a)}\hat{s}_\mu^{(a)}\sqrt{k_\mu}\right] = \exp\left[-\frac{1}{2N}\sum_{i=1}^{N}\sum_{\mu=1}^{M}\sum_{a,b=1}^{n}\hat{s}_\mu^{(a)}\hat{s}_\mu^{(b)}\phi_i^{(a)}\phi_i^{(b)}k_\mu\right]. \tag{4.156}$$

At this level, we can introduce the usual overlap matrix:

$$Q_{ab} \equiv \frac{1}{N}\sum_i\phi_i^{(a)}\phi_i^{(b)} \tag{4.157}$$

and rewrite the resolvent (4.160) as:

$$
R(z) = \lim_{n \to 0} \frac{1}{N} \left(-\frac{2}{n} \right) \frac{\partial}{\partial z} \int_{-\infty}^{+\infty} \prod_{a=1}^{n} \prod_{i=1}^{N} \frac{d\phi_i^{(a)}}{\sqrt{2\pi/i}} \int_{-\infty}^{+\infty} \prod_{a=1}^{n} \prod_{\mu=1}^{M} \frac{d\hat{s}_\mu^{(a)} d\hat{s}_\mu^{(b)}}{2\pi}.
$$

$$
\int_{-\infty}^{+\infty} \prod_{a \leq b} \frac{dQ_{ab} d\hat{Q}_{ab}}{2\pi} \exp\Bigg(-\frac{izN}{2} \sum_{a=1}^{n} Q_{aa} + \frac{i}{2} \sum_{a=1}^{n} \sum_{\mu=1}^{M} (s_\mu^{(a)})^2 +
$$

$$
+ i \sum_{a=1}^{n} \sum_{\mu=1}^{M} s_\mu^{(a)} \hat{s}_\mu^{(a)} - \frac{1}{2} \sum_{a,b=1}^{n} \sum_{\mu=1}^{M} \hat{s}_\mu^{(a)} \hat{s}_\mu^{(b)} Q_{ab} k_\mu + iN \sum_{a \leq b}^{n} Q_{ab} \hat{Q}_{ab} +
$$

$$
- i \sum_{a \leq b}^{n} \hat{Q}_{ab} \sum_{i=1}^{N} \phi_i^{(a)} \phi_i^{(b)} \Bigg).
$$

(4.158)

The integration over the fields ϕ_i gives a factor $\exp\left(\frac{N}{2} \mathrm{Tr} \log Q \right)$. Provided that the stiffness coefficients k_μ and the random part $\xi_i^\mu \xi_j^\mu$ in Eq. (4.142) can be considered uncorrelated for large N, we recover in the SAT phase as well a modified Wishart matrix. In the *annealing* hypothesis, i.e. taking only one replica, we can write the integrand in a more compact way. In other words, we are assuming that the overlap matrix $Q_{ab} = \delta_{ab} R$.

Inserting the variable change term $\exp\left(\frac{N}{2} \mathrm{Tr} \log Q \right)$ in the expression for the resolvent and taking advantage of further simplifications, i.e. separating the terms that depend explicitly on μ, we obtain for the integrand:

$$
\exp\left[-\frac{1}{2}(\lambda + \zeta)R + \frac{1}{2} \log R - \frac{1}{2N} \sum_{\mu=1}^{M} \log(1 - k_\mu R) \right].
$$

(4.159)

where the stiffness distribution is not specified yet. A saddle-point computation allows us to write:

$$
\lambda + \zeta = \frac{1}{R} + \frac{1}{N} \sum_{\mu=1}^{M} \frac{k_\mu}{1 - k_\mu R} = \frac{1}{R} + F'(R)
$$

(4.160)

where

$$
F(R) = -\frac{1}{N} \sum_{\mu=1}^{M} \log(1 - k_\mu R).
$$

(4.161)

An alternative way to derive this equation might be to consider the expectation value of the resolvent in field theory whose cumulants correspond to the correlators. These correlators, once a loop insertion operator is introduced, satisfy loop equations, namely the *Schwinger-Dyson equations* or *Pastur equations* in mathematical jargon [8]. Let us start considering the simplest case in which there is no difference between

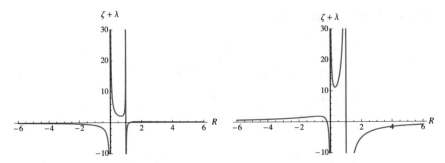

Fig. 4.8 Qualitative behavior of $\tilde{\lambda} = \lambda + \zeta$ as a function of R obtained respectively for $\alpha = 0.2$ and $\alpha = 5$

the coefficients k_μ, namely:

$$\lambda + \zeta = \frac{1}{R} + \frac{\alpha}{1 - R} . \tag{4.162}$$

We show below two qualitative plots reproducing the behavior of $\lambda + \zeta$ as a function of the resolvent, R (Fig. 4.8). As soon as the parameter α becomes greater than one, the plot appears to be visibly different and a translation of the maximum from the positive real axis to the negative part occurs. We can also note that the absolute minimum gets closer to the first pole, which is located in the origin. The points corresponding to a zero first derivative of Eq. (4.162) with respect to R are:

$$R_{min,max} = \frac{1}{1 \pm \sqrt{\alpha}} . \tag{4.163}$$

Anyway, to study the behavior of λ close to $-\zeta$ it looks more convenient to rescale the resolvent as:

$$R = \frac{1}{\lambda + \zeta} H \quad \Rightarrow \quad 1 = \frac{1}{H} + \frac{\alpha}{\lambda + \zeta - H} . \tag{4.164}$$

The points where $\Im[H_0] = 0$ define the edges of the spectrum of H_0 with real solutions only and are obtained imposing the condition:

$$-\frac{1}{H_0}^2 + F''[H_0] = 0 \quad \Rightarrow \quad -\frac{1}{H_0}^2 + \frac{\alpha}{(\tilde{\lambda} - H_0)^2} = 0 , \tag{4.165}$$

where $\tilde{\lambda} = \lambda + \zeta$. Therefore, Eq. (4.165) leads to:

$$\tilde{\lambda}_0 = (1 \pm \sqrt{\alpha}) H_0 . \tag{4.166}$$

By evaluating Eq. (4.164) in H_0 we obtain a quadratic equation whose imaginary part must satisfy the Marchenko–Pastur law [30]:

$$4\tilde{\lambda} - (1 - \alpha + \tilde{\lambda})^2 = (\tilde{\lambda} - \tilde{\lambda}_-)(\tilde{\lambda}_+ - \tilde{\lambda}) \,, \tag{4.167}$$

$$\tilde{\lambda}_-^2 - 2\tilde{\lambda}_-(1 + \alpha) + (1 - \alpha)^2 = 0 \,. \tag{4.168}$$

The second relation yields: $\lambda_- = 1 + \alpha \pm 2\sqrt{\alpha}$ and, because of the condition $\lambda_+ + \lambda_- = 2(1 + \alpha)$, we also obtain $\lambda_+ = (1 \pm \sqrt{\alpha})^2$. We have to pay attention to select the valid solutions according to the plots above. Indeed, the only ones to take into account are:

$$\tilde{\lambda}_\pm = (1 \pm \sqrt{\alpha})^2 \,. \tag{4.169}$$

This simple computation confirms the expected Marchenko–Pastur-like behavior. Anyway, we aim to understand the complicated vibrational spectrum in the whole SAT phase, irrespective of the nature of the stiffness coefficients. The next step will be to consider all different stiffness coefficients whose probability distribution is unknown *a priori*.

4.8.3 Asymptotic Behavior of the Spectral Density in the Jamming Limit

Equation (4.160) cannot be solved exactly. We can nevertheless identify the leading behaviors of the distribution of eigenvalues in the low-frequency regime, provided that the system is analyzed close to jamming. In the RS phase the spectrum is gapped and it becomes gapless when entering the full RSB phase.[9] This justifies our choice to indicate in the following the lower edge of the spectrum as λ_0. Corresponding to the value for which $R = R_0$, the lower edges fulfills the following equations:

$$\lambda_0 + \zeta = \frac{1}{R_0} + F'(R_0) \tag{4.170}$$

$$-\frac{1}{R_0^2} + F''(R_0) = 0 \,, \tag{4.171}$$

which allow to understand how, close to R_0, there is no real linear solution $\lambda - \lambda_0 \propto R - R_0$. Therefore:

$$\lambda - \lambda_0 = \left(\frac{1}{R_0^3} + F'''(R_0) \right) (R - R_0)^2 \,. \tag{4.172}$$

---

[9]We expect the lower edge of the spectrum to be zero in the whole glassy phase, as a consequence of the marginal stability condition.

It is then immediate to see that, for non-zero ζ, the spectral density displays a square-root scaling close to the edge, i.e. $\rho(\lambda) \sim \sqrt{\lambda - \lambda_0}$. From (4.171) one can observe that the condition that the spectral gap λ_0 is positive coincides with the condition that the so-called replicon eigenvalue in the replica solution of the model is positive, as shown in Fig. 4.9. This is known to be positive in the liquid (RS) phase and in the stable (1RSB) phases, while it vanishes in the (full RSB) spin-glass or marginal Gardner glass phases. In [22, 23] it has been shown that close to the jamming line the perceptron enters a Gardner phase, where the replicon is zero and the spectral gap vanishes. Moreover, upon approaching the jamming line $\zeta \to 0$ and $R \to \infty$. These asymptotic behaviors mean that the square-root behavior of the spectrum is modified for $\lambda \sim \zeta$. To better understand this argument we need to study the following integral over the stiffness probability distribution $p(k) = (1 - C)\delta(k) + C P(k)$, where C is the fraction of non-zero forces, approaching 1 at jamming. Let us focus on $F'(R)$:

$$F'(R) = \frac{1}{R} C \int_0^\infty dk\, P(k) \frac{kR}{1 - kR}\ , \qquad (4.173)$$

evaluated for large R. Clearly, the first term in the expansion is $F'[R] \approx C/R$, for $C \to 1$. This term cancels out with the former in (4.155). We thus only need to evaluate the next term in the expansion:

$$F'(R) = \frac{1}{R} C \left[1 - \int_0^\infty dk\, P(k) \frac{1}{1 - kR} \right] . \qquad (4.174)$$

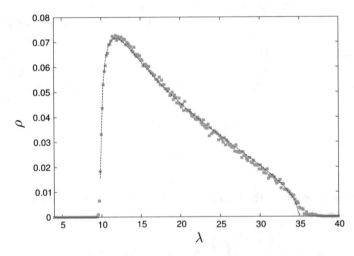

Fig. 4.9 Spectral density of the Hessian matrix in the RS phase. The plot has been reproduced solving numerically the TAP equations for m_i and f_μ, until convergence, and then computing the eigenvalues of the Hessian matrix. The dashed black line is obtained by a Marchenko-Pastur law fit, in remarkable agreement with the data points.

Since the last piece depends critically on the probability $P(k)$ close to the origin, a more detailed analysis is needed in this regime. We recall that the computation of the stiffness coefficient leads to:

$$k_\mu = -[\Phi''(f_\mu) + (1-q)]^{-1} , \qquad (4.175)$$

which, as $q \to 1$, implies that:

$$\Phi''(f_\mu)^{-1} = -\frac{1}{1-q} + \frac{1}{(\sigma - v_\mu)^2} . \qquad (4.176)$$

Then, using the results of Sect. 4.5, we conclude that the stiffness terms roughly coincide with the squared moment of the generalized forces, i.e. $k_\mu \approx f_\mu^2$.

One then needs to know the exact force distribution, which, quoting from the replica solution [21, 22], behaves as a power law $P(f) \sim f^\theta$ with a non-trivial exponent $\theta \sim 0.42311$. Based on these results, we conclude that the integral below has a divergent behavior like:

$$\int_0^\infty dk\, P(k)\frac{1}{k} \approx \int_0^\infty df\, P(f)\frac{1}{f^2} , \qquad (4.177)$$

which rewritten explicitly as a function of the exponent θ becomes:

$$\int df\, P(f)\frac{1}{1+f^2 R} \approx \frac{D}{R^{\frac{1+\theta}{2}}} \qquad (4.178)$$

where D is a constant. Therefore, to leading order the equation for the resolvent reads:

$$\lambda + \zeta = \frac{1-C}{R} + \frac{D}{R^{\frac{3+\theta}{2}}} , \qquad (4.179)$$

from which two different behaviors can be highlighted depending on how much close to the jamming line the system is:

$$\rho(\lambda) \sim \begin{cases} \sqrt{\lambda} & \text{for } \lambda \ll \zeta \\ (\zeta+\lambda)^{-\frac{2}{3+\theta}} & \text{for } \lambda \geq \zeta . \end{cases} \qquad (4.180)$$

These two regimes are well interpolated by:

$$\rho(\lambda) \approx Const.\frac{\sqrt{\lambda}}{(\zeta+\lambda)^{\frac{7+\theta}{2(3+\theta)}}} . \qquad (4.181)$$

We want to say a few words to emphasize the importance of this result. It surprisingly coincides with that found by [18] by analyzing finite-dimensional hard-sphere glasses.[10] This perfect mapping can be immediately understood by expressing $\rho(\lambda)$ in terms of the density of states, namely: $D(\omega) = \rho(\lambda)\frac{d\lambda}{d\omega}$ with $\lambda = \omega^2$.

$$D(\omega)d\omega = \rho(\lambda)d\lambda \, ,$$
$$\rho(\lambda) = D(\omega)\frac{d\omega}{d\lambda}\bigg|_{\omega=\sqrt{\lambda}} . \qquad (4.182)$$

According to [18], we know that the characteristic scaling of the spectral density is $D(\omega) \sim \omega^{-a}$ where $a = (1 - \theta)/(3 + \theta)$. It explicitly depends on the exponent θ of the probability distribution of forces. Expressing the spectral density as a function of the exponent a (or equivalently of θ) we obtain:

$$\rho(\lambda) = \lambda^{-\frac{a}{2}} \frac{1}{2\sqrt{\lambda}} = \frac{1}{2}\lambda^{-\frac{1+a}{2}} \, , \qquad (4.183)$$

$$\rho(\lambda) \sim \lambda^{-\frac{2}{3+\theta}} . \qquad (4.184)$$

The two results are in perfect agreement, irrespective of the dimension of the system: despite the mean-field nature of the perceptron, this model allows us to obtain the same predictions made in finite dimensions.

Notice that the behavior of the spectral density differs from that recovered for the Hessian of the Hamiltonian on the UNSAT side of the transition, where $\rho_{UNSAT}(\lambda) \propto \frac{\sqrt{\lambda}}{\lambda+\zeta}$ and $\zeta \to 0$ at the transition [11, 22]. This discontinuity should not be a surprise, considering that the way to take the zero-temperature limit yields different conditions in the two phases. In the SAT phase, any increase of the energy away from zero is forbidden, vibrations are purely entropic and their amplitude is proportional to the cage size scaling as $\sqrt{1-q}$. While in the SAT phase different replicas can be in different configurations, implying that $q < 1$, in the UNSAT region they can only be in the same absolute minimum, implying that $q = 1$ at $T = 0$. In the second case, the zero-temperature limit must be taken with $q \to 1$ and $\chi = \beta(1 - q)$ fixed.

4.9 Conclusions

In this chapter we introduced a very well-known model in machine learning, the perceptron, both in its seminal derivation proposed by Gardner and in a more recent version. Several analytical simplifications and significant results come from the study of this variant, thanks, in particular, to its peculiar features in the unconventional (non-

---

[10]The authors derived this result using the effective medium theory (EMT), which consists in a mean-field approximation where the disorder is treated in a self-consistent way. More exactly, a random elastic network can be mapped into a regular lattice where each spring constant depends implicitly on the effective frequency. The disorder is determined then by the presence or the absence of a spring in the network.

convex) regime. Indeed, the non-convex perceptron undergoes a critical jamming transition, which enables a direct connection with high-dimensional sphere models. By varying the number of constraints and the number of variables, the perceptron might display two different phases: a SAT phase, corresponding to a hard-sphere regime, and an UNSAT one, which can be mapped into a soft-sphere problem. A one-to-one mapping between this SAT/UNSAT transition and the jamming onset in amorphous solids actually exists.

We proposed an analytical derivation of the TAP free energy, which serves as a coarse-grained functional after integrating out fast degrees of freedom. We developed the computation up to the third order, proving that higher order corrections do not contribute in the jamming limit, as correctly expected according to the isostaticity argument. The only relevant contribution is given by binary interactions, and hence the previous computation at a mean-field level continue to hold. On the other hand, these many-body interactions would provide a non-vanishing contribution either in accounting for finite-size effects or in studying not completely-connected systems, for instance on a random graph with patterns of finite connectivity.

The main outcome of this chapter can be nevertheless considered the definition of a suitable effective potential, which appears as an essential tool to define a Hessian in hard-sphere models and enables us to compute the spectrum of vibrational modes. Even though the Hessian matrix is again a random matrix from a (modified) Wishart ensemble, at frequencies greater than the cut-off frequency ω^*, the spectral density shows a completely different behavior compared to the soft-sphere regime.

References

1. Altieri A (2018) Higher-order corrections to the effective potential close to the jamming transition in the perceptron model. Phys Rev E 97(1):012103
2. Altieri A, Franz S, Parisi G (2016) The jamming transition in high dimension: an analytical study of the TAP equations and the effective thermodynamic potential. J Stat Mech: Theory Exp 2016(9):093301
3. Aspelmeier T, Bray AJ, Moore MA (2004) Complexity of Ising spin glasses. Phys Rev Lett 92(8):087203
4. Baule A et al (2018) Edwards statistical mechanics for jammed granular matter. Rev Mod Phys 90(1):015006
5. Biroli G (1999) Dynamical TAP approach to mean field glassy systems. J Phys A Math Gen 32(48):8365
6. Biroli G, Mézard M (2001) Lattice glass models. Phys Rev Lett 88(2):025501
7. Bolthausen E (2014) An iterative construction of solutions of the TAP equations for the Sherrington-Kirkpatrick model. Commun Math Phys 325(1):333
8. Borot G et al (2011) Large deviations of the maximal eigenvalue of random matrices. J Stat Mech: Theory Exp 2011(11):P11024
9. Brito C, Wyart M (2009) Geometric interpretation of previtrification in hard sphere liquids. J Chem Phys 131(2):024504
10. Brito C, Wyart M (2006) On the rigidity of a hard-sphere glass near random close packing. EPL (Europhys Lett) 76(1):149
11. Charbonneau P et al (2016) Universal non-Debye scaling in the density of states of amorphous solids. Phys Rev Lett 117(4):045503

12. Cover TM (1965) Geometrical and statistical properties of systems of linear inequalities with applications in pattern recognition. IEEE Trans Electron Comput EC-14:326
13. Crisanti A, Horner H, Sommers H-J (1993) The spherical p-spin interaction spin-glass model. Z für Phys B Condens Matter 92(2):257
14. Crisanti A, Sommers H-J (1995) Thouless-Anderson-Palmer approach to the spherical p-spin spin glass model. Journal de Physique I 5(7):805
15. De Almeida JRL, Thouless DJ (1978) Stability of the Sherrington-Kirkpatrick solution of a spin glass model. J Phys A Math Gen 11(5):983
16. De Dominicis C, Martin PC (1964) Stationary entropy principle and renormalization in normal and superfluid systems. II. Diagrammatic formulation. J Math Phys 5(1):31
17. DeGiuli E, Lerner E, Wyart M (2015) Theory of the jamming transition at finite temperature. J Chem Phys 142(16):164503
18. DeGiuli E et al (2014) Force distribution affects vibrational properties in hardsphere glasses. Proc Natl Acad Sci 111(48):17054
19. Edwards SF (1991) The aging of glass forming liquids. Disord Condens Matter Phys, 147
20. Edwards SF, Oakeshott RBS (1989) Theory of powders. Phys A Stat Mech Appl 157(3):1080
21. Franz S, Parisi G (2016) The simplest model of jamming. J Phys A Math Theor 49(14):145001
22. Franz S et al (2015) Universal spectrum of normal modes in low-temperature glasses. Proc Natl Acad Sci 112(47):14539
23. Franz S et al (2017) Universality of the SAT-UNSAT (jamming) threshold in nonconvex continuous constraint satisfaction problems. SciPost Phys 2(3):019
24. Gardner E (1988) The space of interactions in neural network models. J Phys A Math Gen 21(1):257
25. Gardner E, Derrida B (1988) Optimal storage properties of neural network models. J Phys A Math Gen 21(1):271
26. Gendelman O et al (2016) Emergent interparticle interactions in thermal amorphous solids. Phys Rev E 94(5):051001
27. Georges A, Yedidia JS (1991) How to expand around mean-field theory using high-temperature expansions. J Phys A Math Gen 24:2173
28. Hansen J-P, McDonald IR (1990) Theory of simple liquids. Elsevier
29. Janson S, Luczak T, Rucinski A (2011) Random graphs, vol 45. Wiley
30. Marchenko VA, Pastur LA (1967) Distribution of eigenvalues for some sets of random matrices. Mat Sb 114(4):507
31. Mari R, Krzakala F, Kurchan J (2009) Jamming versus Glass Transitions. Phys Rev Lett 103(2):025701
32. Mehta ML (2004) Random matrices, vol 142. Elsevier
33. Mezard M (1989) The space of interactions in neural networks: Gardner's computation with the cavity method. J Phys Math Gen 22(12):2181
34. Montanari A, Mézard M (2009) Information, physics and computation. Oxford University Press
35. Nakanishi K, Takayama H (1997) Mean-field theory for a spin-glass model of neural networks: TAP free energy and the paramagnetic to spin-glass transition. J Phys Math Gen 30(23):8085
36. Parisi G, Mézard M, Virasoro MA (1987) Spin glass theory and beyond. World Scientific Singapore
37. Parisi G, Zamponi F (2010) Mean-field theory of hard sphere glasses and jamming. Rev Mod Phys 82(1):789
38. Plefka T (1982) Convergence condition of the TAP equation for the infinite-ranged Ising spin glass model. J Phys Math Gen 15(6):1971
39. Thouless DJ, Anderson PW, Palmer RG (1977) Solution of 'Solvable model of a spin glass'. Philos Mag 35(3):593
40. Venkatesh SS (1986) Epsilon capacity of neural networks. In: AIP conference proceedings, vol 151. 1. AIP
41. Wyart M (2012) Marginal stability constrains force and pair distributions at random close packing. Phys Rev Lett 109:125502

42. Wyart M (2005) On the rigidity of amorphous solids. Annales de Physique 30:1
43. Wyart M, Nagel SR, Witten TA (2005) Geometric origin of excess lowfrequency vibrational modes in weakly connected amorphous solids. EPL (Europhys Lett) 72(3):486
44. Zamponi F (2010) Mean field theory of spin glasses. arXiv:1008.4844
45. Zdeborová L, Krzakala F (2007) Phase transitions in the coloring of random graphs. Phys Rev E 76(3):031131
46. Zdeborová L, Krzakala F (2016) Statistical physics of inference: thresholds and algorithms. Adv Phys 65(5):453

Chapter 5
Universality Classes: Perceptron Versus Sphere Models

This chapter aims to emphasize the connections between the spherical perceptron and generic sphere models, subject either to hard-core or to a soft potential. The following chapter is structured in two parts. In the first part, we present results already published in [2], based on the generalization of the TAP formalism to soft-sphere systems. In the second part, we consider the analysis of condensation phenomena in mass-transfer models, which was inspired by a stimulating discussion with Emmanuel Trizac and Jacopo Rocchi at LPTMS.

Studying sphere models either in finite or infinite dimensions from first principles can be extremely difficult. One might wonder whether the introduction of auxiliary models, in principle different from the original one but strictly related to it, can allow a more detailed treatment and at the same time efficiently reproduce its main physical properties. A possible setting, which retains a mean-field-like nature, is described as follows. We consider a simple background of three particles: if particle A interacts strongly with two other particles, B and C, these last two have a very weak probability of interacting strongly one with the other. This phenomenon can be explained either by invoking the property for which individual interactions are weak or by referring to the tree-like structure of the network. Models displaying such a structure are very familiar and can be treated with ordinary spin-glass theory tools. Different ways to deal with particle systems in a mean-field-like framework have been proposed over the years. A possible way is to introduce infinite-range interaction models. If the potential itself induces a strong frustration, there is no need to introduce quenched disorder by hand. A second approach is to exploit long but finite-range interactions, analogously to the so-called technique *à la Kac* known in the spin-glass literature. Alternatively, one can also define models on a Bethe lattice [7, 10, 17, 25], which have both a mean-field nature and a quenched disorder. In this framework, by sending the graph connectivity to infinity, one typically recovers the original infinite-dimensional model. A detailed description of models defined on a lattice will be substantiated in Chap. 7 in order to present a suitable perturbative expansion, which is valid even in finite-dimensional systems.

© Springer Nature Switzerland AG 2019
A. Altieri, *Jamming and Glass Transitions*, Springer Theses,
https://doi.org/10.1007/978-3-030-23600-7_5

What we want to describe here is a different approach, according to [18]. We thus introduce the following Hamiltonian:

$$H\left(\{\vec{x}, \vec{A}\}\right) = \sum_{\langle ij \rangle} V(x_i - x_j - A_{ij}) , \qquad (5.1)$$

where V denotes a short-range potential. The particles are arranged in positions $\vec{x}$'s, while the A's are quenched random variables extracted from a distribution $P(|A|)$. Without any loss of generality, we can assume a symmetry rule, namely $A_{ij} = A_{ji}$. The tunable parameter λ, representing the variance of the probability distribution of the A's, assures a proper interpolation between a finite-dimensional model (for $\lambda = 0$ and $P(|A|) = \delta(|A|)$) and a mean-field-like one (for $\lambda \to \infty$). In the liquid phase the mean-field model is solely described by the ideal gas entropic term and the first term in the virial expansion. This simplified behavior is again due to the low probability of having three or more particles interacting simultaneously. For three particles labeled as i, j, k, with diameter σ, one should have:

$$\begin{aligned} |x_i - x_j - A_{ij}| &\sim \sigma \\ |x_j - x_k - A_{jk}| &\sim \sigma \\ |x_k - x_i - A_{ki}| &\sim \sigma \end{aligned} \qquad (5.2)$$

which also implies that $|A_{ij} + A_{jk} + A_{ki}| \sim \sigma$, very unlikely in high dimensions or if the shift is of the same order as the linear size of the system.

Let us start by considering a monodisperse system, described by the following potential:

$$V(u) = \begin{cases} \infty, & \text{for } u > \sigma \\ 0, & \text{otherwise} . \end{cases} \qquad (5.3)$$

According to Eq. (5.1), the partition function reads:

$$Z_{\{A\}} = \int \prod_k dx_k \exp\left(-\sum_{ij} V(x_i - x_j - A_{ij})\right) . \qquad (5.4)$$

In the liquid phase the annealed approximation safely applies, allowing us to write the entropy as:

$$S = -\overline{\ln Z_{\{A\}}} = -\ln \overline{Z_{\{A\}}} . \qquad (5.5)$$

The computation simply reduces to the study of the averaged partition function:

$$\overline{Z_{\{A\}}} = \frac{1}{N!} \int \prod_{mn} P(A_{mn}) dA_{mn} \int \prod_k dx_k \exp\left(-\sum_{ij} V(x_i - x_j - A_{ij})\right) , \qquad (5.6)$$

where the factorial is needed for a correct Mayer expansion [14]. More conveniently the computation can be performed in the grand-canonical ensemble, which requires the introduction of a grand-canonical partition function:

$$\overline{\mathcal{Z}}_{GC} = \sum_N z^N \overline{Z_{\{A\}}} \, , \tag{5.7}$$

where $z = e^\mu$ is the fugacity, which depends on the chemical potential μ. In the grand-canonical ensemble, the partition function can be rewritten as:

$$\overline{Z_{\{A\}}} = \int \prod_k dx_k \prod_{ij} \left[\overline{f} \left(x_i - x_j \right) + 1 \right] \, , \tag{5.8}$$

depending on the Mayer function, which in the annealed approximation reads:

$$\overline{f}(x - y) = \int dA P(A) \left[\exp\left(-V(x - y - A) \right) - 1 \right] = - \int dA P(A) \chi \left(|x - y - A| \right) \, . \tag{5.9}$$

We indicated in the expression above the step function as $\chi(r)$ equal to one if $r < \sigma$ and zero otherwise. At this point, the grand-canonical potential defined as $\mathcal{G} = \ln \overline{\mathcal{Z}}_{GC}$ can be expressed via a diagrammatic expansion, taking into account connected diagrams only. Indeed, taking the logarithm of the expansion automatically removes disconnected diagrams.

Each vertex has a contribution z, while each link is associated with a Mayer function $\overline{f}$. According to whether the order of the vertices is greater or smaller than the order of the bonds, the corresponding diagram provides either a finite or a vanishing contribution in the expansion. Hence, in the annealed approximation, the entropy of the model can be written as a sum of the entropic logarithmic term and the first leading term of the virial expansion [14, 23], both functions of the density profile $\rho(x)$:

$$S_{an}[\rho(x)] = - \int dx \rho(x) \left[\ln \rho(x) - 1 \right] + \frac{1}{2} \int dx dy \rho(x) \rho(y) \overline{f}(x - y) + N \ln N \, , \tag{5.10}$$

where the last term is justified in terms of the not truly indistinguishable nature of the particles. The integral (5.9) corresponds to the volume of a sphere of diameter equal to 1 and position $x - y$. It thus gives a contribution:

$$\overline{f}(x - y) = -v_d/V \, . \tag{5.11}$$

Therefore in the Mayer expansion each diagram with n vertices and m bonds contributes as $V^n z^z (-v_d/V)^m = V^{n-m} z^n (-v_d)^m$. It is immediate to see that if $m \geq n$ the diagram does not provide any contribution in the thermodynamic limit, while in the opposite case, for $m < n$, it does. The first possibility, for $m \geq n$, corresponds to have loopy structures, which are vanishingly small when $d \to \infty$ or in generic mean-field models. In the second case, because $m = n - 1$, the surviving diagrams have a tree-like structure.[1]

An important point to remark is that the equation of state valid in the liquid phase and derivable from Eq. (5.10) exactly coincides with that derived for hard spheres in infinite dimension [13, 22]. The formal analogy between these two models as $\lambda \to \infty$ and $d \to \infty$ suggests using this so-called Mari-Kurchan model as a reasonable starting point in several cases. The fact the only tree-like diagrams contribute to the above expansion can be also proven in the Thouless-Anderson-Palmer formalism, as we did for the perceptron model and we repeat here in sphere models.

5.1   TAP Formalism Generalized to Sphere Models

In Chap. 4 we have largely explained the perceptron model and its connection with the jamming paradigm. The main goal of this Section is to highlight the centrality of the perceptron model, safely generalizable to sphere models in the limit of large dimensions. The underlying assumption is to consider the gap variable as a function of two particle positions, that is the index $\mu \to (\alpha, \beta)$ now codes for particle pairs with diameter σ.

In the following, we will look at arrangements of M spheres on the surface of a N-dimensional hypersphere of radius R. This leads to a simpler analysis than spheres in Euclidean space, but it should be equivalent to it in the limit $\sigma/R \to 0$.

In fact, there are different possible high-dimensional limits of this model: the natural one [15, 16] consists in taking $R, M \to \infty$ and then $N \to \infty$, for fixed reduced packing fraction $\hat{\phi} = 2^N \frac{M}{N} (\sigma/2R)^{N-1}$; the second one instead considers $N \to \infty$, while $R = \sqrt{N}$ and $M = \alpha N$.

The generalization of the perceptron model to high-dimensional spheres can be easily visualized by introducing $M = \alpha N$ particles in positions x_α, $(\alpha = 1, ..., M)$ arranged over a N-dimensional sphere of radius $R = \sqrt{N}$. The normalization, i.e. $x_\alpha^2 = N$, continues to hold, but the gap variables now take the form:

$$h_{\alpha\beta} = \frac{x_\alpha \cdot x_\beta}{\sqrt{N}} - \sigma .\qquad(5.12)$$

The crucial point is to work in a regime where each particle effectively interacts with $O(N)$ other particles, because both of the aforementioned regimes have this property. To perform a suitable Plefka-like expansion we introduce the following Hamiltonian:

---

[1]This is a conventional relation in diagrammatic theory, which connects the number of loops in a given diagrams with the number of internal lines and vertices, i.e. $L = I - V + 1$. Since in the model described above [18] the only relevant diagrams have a tree-like structure, that is $L = 0$, we exactly recover the relation $I = V - 1$.

$$H_{\text{eff}} = i \sum_{\langle \alpha, \beta \rangle} \frac{\hat{h}_{\alpha\beta} x_\alpha \cdot x_\beta}{\sqrt{N}} \,, \tag{5.13}$$

and by making use of the same treatment as for the perceptron, we obtain an expansion in terms of the first two moments of the free-energy functional $\Gamma(\vec{m}, \vec{f})$:

$$\frac{\partial \Gamma}{\partial \eta} = -\langle H_{\text{eff}} \rangle = -\sum_{\langle \alpha\beta \rangle} \frac{f_{\alpha\beta} m_\alpha \cdot m_\beta}{\sqrt{N}} \,, \tag{5.14}$$

$$\frac{\partial^2 \Gamma}{\partial \eta^2} = -\left[\langle H_{\text{eff}}^2 \rangle - \langle H_{\text{eff}} \rangle^2 - \langle H_{\text{eff}} \sum_i (s_i - m_i) \frac{\partial \langle H_{\text{eff}} \rangle}{\partial m_i} \rangle \right] \tag{5.15}$$

where, for compactness we indicated with s_i both types of variables, *i.e.* positions and contact forces. We normalize the parameters of the model as follows:

$$q = \frac{1}{NM} \sum_{i,\alpha} (m_i^\alpha)^2 = \frac{1}{M} \sum_{\alpha=1}^{M} \frac{m^\alpha \cdot m^\alpha}{N} \,, \quad r = -\frac{1}{MN} \sum_{\alpha\beta} f_{\alpha\beta}^2 \,, \quad \tilde{r} = -\frac{1}{MN} \sum_{\alpha\beta} \langle \hat{h}_{\alpha\beta}^2 \rangle \,. \tag{5.16}$$

Note that this normalization choice immediately suggests that there are $O(N)$ forces $f_{\alpha\beta}$ for each sphere.

Let us focus on the second-order term of the expansion. The first contribution in Eq. (5.15), namely the connected part of the Hamiltonian, yields:

$$\langle H_{\text{eff}}^2 \rangle - \langle H_{\text{eff}} \rangle^2 = -\frac{1}{N} \sum \left(\langle \hat{h}_{\alpha\beta} \hat{h}_{\gamma\delta} \rangle \langle x_i^\alpha x_i^\beta x_j^\gamma x_j^\delta \rangle \right)_c = -MN(\tilde{r} - rq^2) \,, \tag{5.17}$$

where the only surviving terms are those with $(\alpha\beta) = (\gamma\delta)$. This is exactly the same argument that was discussed in Chap. 4 and for which we refer the interested reader to the Appendix.

Using the definition $H_{eff} = i\hat{h}_{\alpha\beta} h_{\alpha\beta}$, one can easily see that $\langle H_{eff}^2 \rangle = \tilde{r}\langle h^2 \rangle = \tilde{r}$ and $\langle H_{eff} \rangle^2 = r\langle h \rangle^2 = rq^2$. Similarly, the second part in Eq. (5.15) can be written as:

$$\sum_{\langle \alpha\beta, \gamma\delta \rangle} \sum_{ij} \langle \frac{i\hat{h}_{\alpha\beta} x_i^\alpha x_i^\beta}{\sqrt{N}} \left[\frac{(x_j^\gamma - m_j^\gamma) f_{\gamma\delta} m_j^\delta}{\sqrt{N}} + \frac{(i\hat{h}_{\gamma\delta} - f_{\gamma\delta}) m_j^\gamma m_j^\delta}{\sqrt{N}} \right] \rangle =$$
$$= MNq(1 - q)r + MN(\tilde{r} - r)q^2 \,, \tag{5.18}$$

In a few passages we manage to get a generic expression for the effective potential in sphere models, where the degree of softness/hardness is actually tuned by the parameter σ.

$$\Gamma(\vec{m}, \vec{f}) = -\frac{MN}{2} \log(1 - q) + \sum_{\alpha\beta} \Phi(f_{\alpha\beta}) - \sum_{\alpha\beta} \frac{f_{\alpha\beta} m_\alpha \cdot m_\beta}{\sqrt{N}} +$$

$$+ \frac{MN}{2} \left[(\tilde{r} - rq^2) + rq(1 - q) + q^2(\tilde{r} - r) \right] . \tag{5.19}$$

$\Phi(f)$ is the same functional as for the perceptron, *i.e.*:

$$\Phi(f) = \min_{\tilde{v}} \left[f\tilde{v} - \log \int \frac{dh d\hat{h}}{2\pi} e^{\frac{\beta}{2} h^2 \theta(-h) - i\hat{h}(h+\sigma) + iv\hat{h} - \frac{\tilde{b}}{2}(\hat{h}^2 - \tilde{r})} \right] \tag{5.20}$$

where now the only difference is due to its double index dependence. As before, $\tilde{v}$ is the Lagrange multiplier associated with the force and $\tilde{b}$ an additional multiplier introduced to enforce the average value of $\hat{h}_{\alpha\beta}^2$. We also checked that third-order $O(\eta^3)$ terms in Eq. (5.19) provide a sub-leading contribution compared to the first two ones, as correctly expected from the structure of this model. Since only the first two moments are the relevant ones in the expansion—that implies a Gaussian distribution of the random components $\xi^{\alpha\beta}$'s—one can verify that the same expansion would work in an equivalent disordered model where the scalar products in the definition of the gaps would be substituted by random couplings:

$$h_{\alpha,\beta}(x) = \frac{1}{N} \sum_{ij} \xi_{ij}^{\alpha\beta} x_i^\alpha x_j^\beta - \sigma \tag{5.21}$$

and the components $\xi_{ij}^{\alpha\beta}$ would be chosen as independent, variance one Gaussian variables. This is a clear signature of the equivalence of self-generated disorder models and models with quenched disorder in high dimension [8, 11, 18, 19]. We shall stress more this concept in the following, as it is fundamental for the purposes of this thesis and of several related works. We refer the reader to Sect. 5.3, where a detailed computation in presence of random couplings is proposed.

5.2   Optimal Packing of Polydisperse Hard Spheres in Finite and Infinite Dimensions

In the following we shall present an alternative computation based on the replica method, both in the annealed and the quenched case, which allows us to have a complete overview, beyond the TAP formalism. Our study is motivated by the possibility to observe and then to consistently describe condensation phenomena in stochastic models of *mass* transport [12, 20, 28]. One can consider, for instance, a lattice model on which the mass is stochastically transferred between sites, provided some specific dynamical rules. In a vast class of complex systems—such as granular clustering [27], network rewirings [3], traffic flow [21], non-neutral birth-death processes in ecosystems [5, 6]—one might observe this kind of phenomenon, in principle very close to Bose-Einstein condensation. The only difference is that in the former the transition to a *condensate* takes place in real space without any limitation due to the dimension. Other examples are wealth condensation in macroeconomics [9] and web networks [4], where a single node or hub can capture a macroscopic fraction of links in the network.

Beyond various applications, we aim to focus on glassy systems and achieve a better understanding of the behavior of polydisperse assemblies of hard particles. Let us first see why this problem is so interesting.

It has been shown that the steady-state single-site mass distribution exhibits a signature of such a transition above a critical value of the density of particles in the system [12]. The particle size distribution changes from an exponential decay at low density to a power-law plus a single-peaked behavior in the denser liquid phase. One can identify a critical density at which the decay becomes slower and slower, well-described by a power law or sometimes a stretched exponential. This is a peculiar feature of a critical fluid phase. A more interesting phenomenon is the emergence of a bump above the critical density. Physically, this means that a single lattice site incorporates the excess mass which cannot be included in the normalized distribution. Theory and simulations agree on the emergence of an interesting physical phase at a rather low volume fraction approximately equal to 0.26, above that the Percus-Yevick equations of state [24] are no longer valid. The appearance of a macroscopic aggregate invalidates the treatment based on the simplest liquid theory. In this spirit, a connection with the Bose-Einstein condensation was first suggested in [28].

Till now the problem has been solved analytically only in low dimensions, with a special attention on a one-dimensional closed chain [12] where the total volume of the particles and the linear size L of the lattice are conserved:

$$\sum_i v_i = V , \qquad \sum_i x_i = L . \qquad (5.22)$$

The simplest model that one can consider consists of hard p-spheres subject to a stochastic dynamics according to which diffusion and volume exchange are allowed. The particles are subject to hard-core interactions, meaning that the distance x_i between the i-th sphere and the $(i + 1)$-th sphere must be greater than $l_i = v_i^{1/p}$, where v_i is the sphere volume. The condensation transition can be proven to be a direct consequence of the constraints in configuration space (see Fig. 5.1).

The model just described turns out to be equivalent to a system of N polydisperse hard rods constrained to a ring of total length L, where each rod has diameter l_i and the gap between two nearest rods g_i must be positive or zero for each degree of freedom i. In the large p limit, the model essentially maps onto a monodisperse problem as the diameter l_i goes to 1 and is no longer dependent on the label i.

One can analyse the problem in the canonical ensemble defined by N hard rods at temperature T and total ring length L. The packing fraction is $\eta = L_p/L = \langle l \rangle \rho$, where $\rho = N/L$ is the density and $\langle l \rangle$ the average rod size. In this case the free energy functional can be written as a sum of an ideal term and an excess contribution, namely:

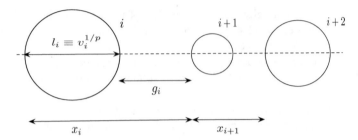

Fig. 5.1 Schematic illustration of three p-spheres on a line with inter-particle distance x_i and gap g_i. The simplest cases can be obtained for $p = 2$ (disks) and $p = 3$ (spheres). The distance between particles i and $i + 1$, x_i, has to be greater than the diameter $l_i = v_i^{1/p}$, where v_i is the particle volume. According to whether the value of p is greater or smaller than one, one can observe a condensation transition or not. Figure taken from [12]

$$\beta \mathcal{F}[W] = N \int dl\, W(l) \left[\ln \left(\Lambda^2 \rho W(l) j_l(l) \right) - 1 \right] + \beta \mathcal{F}_{\text{ex}} \qquad (5.23)$$

where Λ is only a numerical prefactor, irrelevant for scaling purposes, whereas $W(l)$ is the probability distribution associated with length l. The second piece, representing the excess free energy, can be exactly computed, *i.e.* $\beta \mathcal{F}_{\text{ex}} = -\ln(1 - \eta)$ [26]. The factor $j_l(l)$ ensures that different choices of particle labeling lead to the same optimal distribution $W^*(l)$.

The next step consists in minimizing the functional $\mathcal{F}[W]$ provided the two constraints on the total length and the p-moment. The minimization with respect to W allows us to determine the optimal size distribution W^*. Any infinitesimal change in W^* has no effect on $\mathcal{F}[W]$.

To determine the optimal volume distribution the authors [12] considered two separate steps: (i) increasing the length of the rod by an amount δl_0, namely $l_0 \to l_0 + \delta l_0$; (ii) rescaling all particles to guarantee the p-th moment conservation. After these two steps the corresponding change in the distribution is given by:

$$\delta W = \frac{1}{N} \left[\delta(l - l_0 - \delta l_0) - \delta(l - l_0) \right] + (1 - \alpha) \frac{d\, [lW]}{dl} . \qquad (5.24)$$

At this point one can write the variation of the ideal and the excess parts of the free-energy functional, which, together with the condition that $\delta F = 0$ for $W = W^*$, enables to prove that the optimal distribution as a function of l reads [12]:

$$W^*(l) = A l^{p-1} \exp\left\{ -\beta Pl + \left(\frac{\beta P_{\text{ex}} \langle l \rangle}{p\eta} - 1 \right) \frac{l^p}{\langle l^p \rangle} \right\} . \qquad (5.25)$$

Additional parameters are the normalization constant A and the pressure, defined as:

$$\beta P = \frac{\rho}{1 - \eta} \ . \tag{5.26}$$

From that, the excess pressure fulfills the identity $P_{\text{ex}} = P\eta$. The packing fraction η and the density are related one to the other by the relation $\eta = \rho\langle l\rangle$. In the low-density regime, when the density is vanishingly small, $P \to 0$ and $\beta P_{\text{ext}}/\eta \to 0$, which imply an exponential distribution for v:

$$W^*(l) \propto l^{p-1} \exp\left(-\frac{l^p}{\langle l^p\rangle}\right) \ . \tag{5.27}$$

By contrast, in the generic case, the distribution (5.25) can be re-expressed as a function of the critical packing fraction η_c:

$$W^*(l) = A l^{p-1} \exp\left[-\frac{\rho l}{1 - \eta} + \left(\frac{\eta - \eta_c}{\eta_c(1 - \eta)}\right)\frac{l^p}{\langle l^p\rangle}\right], \tag{5.28}$$

where η_c corresponds to $p/(1 + p)$. Two different situations can occur depending on the value of p: if $p \le 1$ the distribution is normalizable $\forall\eta$, whereas if $p > 1$ and $\eta > \eta_c$ the distribution develops a divergent trend at large l. A divergent distribution is a signature of the appearance of a bump, in addition to either a decaying power-law or stretched exponential behavior. This macroscopic agglomerate of particles accounts for the extra mass that cannot be embedded in the well-normalized distribution.

5.3 Condensation in High Dimensions?

The more you get the more you want. One might wonder whether a similar phenomenon can characterize high-dimensional fully-connected models for which an exact solution exists. The main goal of this section is to study a polydisperse system of hard spheres in high dimensions using the formalism introduced at the beginning of this chapter.

Then, the Hamiltonian of the system can be written as a function of the random gaps $h_{\alpha\beta}$:

$$H = \sum_{\alpha\neq\beta} v\left(h_{\alpha\beta}\right), \tag{5.29}$$

where $v(h_{ab})$ is a generic potential, constrained to be positive in the soft-sphere regime, if $h_{ab} < 0$ and zero in the hard-sphere phase, if $h_{ab} > 0$. The gaps $h_{\alpha\beta}$ are functions of both the system configurations x_α and the variables σ_α, where the index $\alpha = 1, ...M$ refers to the particles, *i.e.*:

$$h_{\alpha\beta} = \frac{1}{N}\sum_{ij} \xi_{ij}^{\alpha\beta} x_i^\alpha x_j^\beta - \sigma_\alpha - \sigma_\beta \ . \tag{5.30}$$

Differently than Eq. (5.12), we aim to study the model in the presence of random couplings. The x_i^α's are defined on the N-dimensional sphere, such that $\sum_{i=1}^{N}(x_i^\alpha)^2 = N$, $\forall \alpha = 1, \ldots, M$. They actually play a dynamical role in the model. By contrast, the $\xi_{ij}^{\alpha\beta}$ are quenched random variables, distributed according to a normal distribution $\mathcal{N}(0, 1)$.

In the following we discuss two cases. We first treat the σ_α's as quenched variables and perform all the computations in this hypothesis. Then, we assume that the σ_α's are annealed variables. In the first case we consider the free energy averaged over both $\vec{\xi}$ and $\vec{\sigma}$:

$$F_Q = -T\,\mathbb{E}_{\vec{\xi},\vec{\sigma}}\,\log[Z_{\vec{\xi},\vec{\sigma}}]\,, \tag{5.31}$$

while in the second case we compute the free energy averaged over $\vec{\xi}$ only:

$$F_A = -T\,\mathbb{E}_{\vec{\xi}}\,\log[Z_{\vec{\xi}}]\,. \tag{5.32}$$

In both cases we make use of the replica trick identity:

$$\lim_{n\to 0}\partial_n\overline{Z^n} = \overline{\log Z} \tag{5.33}$$

where ∂_n is nothing but the derivative with respect to n, with the limit $n \to 0$ taken at the end.

5.3.1   Quenched Computation

Let us start our discussion in the case of quenched $\vec{\sigma}$'s. For a given realization of $\vec{\xi}$ and $\vec{\sigma}$, the partition function of the model reads:

$$Z_{\vec{\xi},\vec{\sigma}} = \int \prod_{\alpha,i}dx_i^\alpha \prod_\alpha \delta\left(\sum_i(x_i^\alpha)^2 - N\right)\int \prod_{\alpha\neq\beta}dh_{\alpha\beta}e^{-\beta\sum_{\alpha\neq\beta}v(h_{\alpha\beta})}\,. \tag{5.34}$$

Then we introduce a Lagrange multiplier $i\hat{h}_{\alpha\beta}$ to enforce the gap value:

$$Z_{\vec{\xi},\vec{\sigma}} = \int \prod_{\alpha,i}dx_i^\alpha \prod_\alpha \delta\left(\sum_i(x_i^\alpha)^2 - N\right)\prod_{\alpha\neq\beta}\int \frac{dh_{\alpha\beta}d\hat{h}_{\alpha\beta}}{2\pi}e^{i\hat{h}_{\alpha\beta}\left(h_{\alpha\beta}-\frac{1}{N}\sum_{ij}\xi_{ij}^{\alpha\beta}x_i^\alpha x_i^\beta+\sigma_\alpha+\sigma_\beta\right)}\,.$$
$$\tag{5.35}$$

In the expression above and everywhere in the following, the integral over $h_{\alpha\beta}$ is extended over $(0, \infty)$, so that we may drop the dependence on $v(h_{\alpha,\beta})$. After replicating the system n times, we can perform the average over $\vec{\xi}$:

$$
\mathbb{E}_{\vec{\xi}}\left[Z_{\vec{\xi},\vec{\sigma}}^{n}\right] = \prod_{\alpha\neq\beta}\prod_{ij}\int d\xi_{ij}^{\alpha\beta}e^{-\frac{1}{2}(\xi_{ij}^{\alpha\beta})^2-i\xi_{ij}^{\alpha\beta}\sum_a\hat{h}_{\alpha\beta}^{(a)}\frac{1}{N}x_i^{(a)}x_j^{(a)}} = e^{-\frac{1}{2}\sum_{\alpha\neq\beta}\sum_{ab}\hat{h}_{\alpha\beta}^{(a)}\hat{h}_{\alpha\beta}^{(b)}Q_{ab}^{\alpha}Q_{ab}^{\beta}}
$$

$$(5.36)$$

where we used latin letters for the replica indices. To write the expression above we implicitly defined the overlap matrix as:

$$
Q_{ab}^{\alpha} = \frac{1}{N}\sum_{i=1}^{N}x_i^{\alpha,(a)}x_i^{\alpha,(b)} \qquad \forall \ a\neq b. \tag{5.37}
$$

For $a = b$ we may define $Q_{aa}^{\alpha} = 1 \ \forall\alpha = 1,\ldots,M$ because of the spherical constraint. According to this definition, we may introduce a delta-function in the partition function, which yields:

$$
\mathbb{E}_{\vec{\xi}}\left[Z_{\vec{\xi},\vec{\sigma}}^{n}\right] = \int \prod_{\alpha,i}\prod_{a}dx_i^{\alpha,(a)}\prod_{\alpha}\prod_{ab}\int dQ_{ab}^{\alpha}\delta\left(\sum_i x_i^{\alpha,(a)}x_i^{\alpha,(b)} - NQ_{ab}^{\alpha}\right)\cdot
$$

$$
\int\prod_{\alpha\neq\beta}\prod_a \frac{dh_{\alpha\beta}^{(a)}d\hat{h}_{\alpha\beta}^{(a)}}{2\pi}e^{i\sum_{\alpha\neq\beta}\sum_a\hat{h}_{\alpha\beta}^{(a)}\left(h_{\alpha\beta}^a+\sigma_\alpha+\sigma_\beta\right)-\frac{1}{2}\sum_{\alpha\neq\beta}\sum_{ab}\hat{h}_{\alpha\beta}^{(a)}\hat{h}_{\alpha\beta}^{(b)}Q_{ab}^{\alpha}Q_{ab}^{\beta}}
$$

$$
= \int\prod_{\alpha,i}\prod_a dx_i^{\alpha,(a)}\prod_{\alpha}\prod_{ab}\int\frac{dQ_{ab}^{\alpha}d\hat{Q}_{ab}^{\alpha}}{2\pi}e^{\hat{Q}_{ab}^{\alpha}\left(\sum_i x_i^{\alpha,(a)}x_i^{\alpha,(b)}-NQ_{ab}^{\alpha}\right)}\cdot
$$

$$
\int\prod_{\alpha\neq\beta}\prod_a \frac{dh_{\alpha\beta}^{(a)}d\hat{h}_{\alpha\beta}^{(a)}}{2\pi}e^{i\sum_{\alpha\neq\beta}\sum_a\hat{h}_{\alpha\beta}^{(a)}\left(h_{\alpha\beta}^{(a)}+\sigma_\alpha+\sigma_\beta\right)-\frac{1}{2}\sum_{\alpha\neq\beta}\sum_{ab}\hat{h}_{\alpha\beta}^{(a)}\hat{h}_{\alpha\beta}^{(b)}Q_{ab}^{\alpha}Q_{ab}^{\beta}}
$$

$$(5.38)$$

The integration over $\vec{x}$ then leads to:

$$
\int_{-\infty}^{\infty}\prod_{\alpha=1}^{M}\prod_{i=1}^{N}\prod_{a=1}^{n}dx_i^{\alpha,(a)}e^{-\sum_{ab}^{n}\hat{Q}_{ab}^{\alpha}\sum_{i=1}^{N}x_i^{\alpha,(a)}x_i^{\alpha,(b)}} = \prod_{\alpha=1}^{M}\left(\sqrt{\frac{(2\pi)^n}{\det(2\hat{Q}^\alpha)}}\right)^{N}
$$

$$
\propto e^{-\frac{N}{2}\sum_{\alpha=1}^{M}\log\det\left(2\hat{Q}^\alpha\right)}.
$$

$$(5.39)$$

A fundamental missing piece is given by the following equality [1]:

$$
\frac{\partial}{\partial M_{ab}}\log\det M = [M^{-1}]_{ba}. \tag{5.40}
$$

Now we can take advantage of the saddle-point method over the parameters $\hat{Q}_{ab}^{\alpha}$ to extrapolate the relation between the two conjugated overlaps, *i.e.* $2\hat{Q}_{ab}^{\alpha} = \left[(Q^\alpha)^{-1}\right]_{ab}$. Once irrelevant prefactors are neglected, the final expression of the partition function reads:

$$\mathbb{E}_{\vec{\xi}}\left[Z^n_{\vec{\xi},\vec{\sigma}}\right] = \int \prod_\alpha \prod_{a\neq b} dQ^\alpha_{ab} e^{\frac{N}{2}\sum_\alpha \log \det Q^\alpha}.$$

$$\int \prod_{\alpha\neq\beta}\prod_a \frac{dh^{(a)}_{\alpha\beta}d\hat{h}^{(a)}_{\alpha\beta}}{2\pi} e^{i\sum_{\alpha\neq\beta}\sum_a \hat{h}^{(a)}_{\alpha\beta}\left(h^a_{\alpha\beta}+\sigma_\alpha+\sigma_\beta\right)-\frac{1}{2}\sum_{\alpha\neq\beta}\sum_{ab}\hat{h}^{(a)}_{\alpha\beta}\hat{h}^{(b)}_{\alpha\beta}Q^\alpha_{ab}Q^\beta_{ab}}.$$

$$(5.41)$$

To decouple replicas in the last line we make use of the RS ansatz, which allows us to write the overlap matrix as: $Q^\alpha_{ab} = \delta_{ab} + q^\alpha(1-\delta_{ab})$, according to which we can rewrite the quadratic term as a function of the overlap matrix, namely:

$$-\frac{1}{2}\sum_{\alpha\neq\beta}\sum_{ab}\hat{h}^{(a)}_{\alpha\beta}\hat{h}^{(b)}_{\alpha\beta}Q^\alpha_{ab}Q^\beta_{ab} = -\frac{1}{2}\sum_{\alpha\neq\beta}\left[\sum_a\left(\hat{h}^{(a)}_{\alpha\beta}\right)^2 + \sum_{ab}\hat{h}^{(a)}_{\alpha\beta}\hat{h}^{(b)}_{\alpha\beta}q^\alpha q^\beta\right]=$$

$$=-\frac{1}{2}\sum_{\alpha\neq\beta}\left[(1-q^\alpha q^\beta)\sum_a(\hat{h}^{(a)}_{\alpha\beta})^2 + q^\alpha q^\beta\left(\sum_a\hat{h}^{(a)}_{\alpha\beta}\right)^2\right].$$

$$(5.42)$$

At this point, we introduce an auxiliary variable z with zero mean and unit variance, so that Eq. (5.41) becomes:

$$\mathbb{E}_{\vec{\xi}}\left[Z^n_{\vec{\xi},\vec{\sigma}}\right] = \int \prod_\alpha dq^\alpha e^{\frac{N}{2}\sum_\alpha \log\det Q^\alpha}\prod_{\alpha\neq\beta}\int \mathcal{D}z \int \prod_a \frac{dh^{(a)}_{\alpha\beta}d\hat{h}^{(a)}_{\alpha\beta}}{2\pi}.$$

$$e^{i\sum_a \hat{h}^{(a)}_{\alpha\beta}\left(h^{(a)}_{\alpha\beta}+\sigma_\alpha+\sigma_\beta\right)-\frac{1}{2}(1-q^\alpha q^\beta)\sum_a(\hat{h}^{(a)}_{\alpha\beta})^2-iz\sqrt{q^\alpha q^\beta}\sum_a \hat{h}^{(a)}_{\alpha\beta}} = \qquad (5.43)$$

$$=\int \prod_\alpha dq^\alpha e^{\frac{N}{2}\sum_\alpha \log\det Q^\alpha}\prod_{\alpha\neq\beta}\int \mathcal{D}z\ \mathcal{I}^n$$

where in the last line we defined the integral[2]:

$$\mathcal{I} = \int_0^\infty \frac{dh_{\alpha\beta}}{\sqrt{2\pi(1-q^\alpha q^\beta)}}e^{-\frac{\left(h_{\alpha\beta}+\sigma_\alpha+\sigma_\beta-z\sqrt{q^\alpha q^\beta}\right)^2}{2(1-q^\alpha q^\beta)}}. \qquad (5.44)$$

We can further simplify the expression above, once by noticing that:

$$\det Q^\alpha = (1-q^\alpha)^{n-1}[1+(n-1)q^\alpha]. \qquad (5.45)$$

The averaged partition function eventually reads:

---

[2]We used the compact notation for the gap variables $h^{(a)}_{\alpha\beta} = h_{\alpha\beta}, \forall a$.

$$\mathbb{E}_{\vec{\xi}}\left[Z_{\vec{\xi},\vec{\sigma}}^n\right] = \int \prod_\alpha dq^\alpha e^{\frac{N}{2}\sum_\alpha \{(n-1)\log(1-q^\alpha)+\log[1+(n-1)q^\alpha]\}}$$

$$\prod_{\alpha\neq\beta} \int \mathcal{D}z \left[H\left(\frac{\sigma_\alpha + \sigma_\beta - \sqrt{q^\alpha q^\beta} z}{\sqrt{1-q^\alpha q^\beta}}\right)\right]^n, \tag{5.46}$$

where, again, $H(x)$ stands for the complementary error function. For a generic observable A and a given distribution $W(\sigma_\alpha)$ of σ_α we define the following average:

$$\mathbb{E}_{\vec{\sigma}}\left[A(\vec{\sigma})\right] = \int \prod_\alpha dW(\sigma_\alpha)\delta\left(\sum_\alpha \sigma_\alpha^p - \Sigma\right) A(\vec{\sigma}). \tag{5.47}$$

The expression of the quenched free energy can be obtained by averaging over $\vec{\sigma}$, differentiating with respect to n and then taking the limit $n \to 0$. This procedure will lead to:

$$F_Q = -\frac{1}{\beta N}\left[\sum_{\alpha\beta} \int \mathcal{D}z \log H\left(\frac{\sigma_\alpha + \sigma_\beta - \sqrt{q^\alpha q^\beta} z}{\sqrt{1-q^\alpha q^\beta}}\right) + \frac{N}{2}\sum_\alpha \left(\log(1-q^\alpha) + \frac{q^\alpha}{1-q^\alpha}\right)\right] \tag{5.48}$$

where we made use of the following equivalence: $\lim_{n\to 0} \int \mathcal{D}z [g(z)]^n = \lim_{n\to 0} e^{n\int \mathcal{D}z \log g(z)}$.

5.3.2 Annealed Computation

In this section we define the $\vec{\sigma}$'s as annealed variables. For a given realization of the disorder $\vec{\xi}$, the partition function of the model can be written as:

$$Z_{\vec{\xi}} = \int \prod_{\alpha,i} dx_i^\alpha \prod_\alpha \delta\left(\sum_i (x_i^\alpha)^2 - N\right) \int \prod_\alpha dW(\sigma_\alpha)\delta\left(\sum_\alpha \sigma_\alpha^p - \Sigma\right)\cdot$$

$$\int \prod_{\alpha\beta} dh_{\alpha\beta} e^{-\beta\sum_{\alpha\neq\beta} v(h_{\alpha\beta})}. \tag{5.49}$$

We can apply the same scheme as before but now paying attention to the fact that we need to replicate the $\vec{\sigma}$'s as well. Hence, the annealed partition function reads:

$$\mathbb{E}_{\vec{\xi}}\left[Z_{\vec{\xi}}^n\right] = \int \prod_\alpha dq^\alpha e^{\frac{N}{2}\sum_\alpha \{(n-1)\log(1-q^\alpha)+\log[1+(n-1)q^\alpha]\}}$$

$$\left(\int \prod_\alpha dW(\sigma_\alpha)\delta\left(\sum_\alpha \sigma_\alpha^p - \Sigma\right) e^{\sum_{\alpha\neq\beta} f(\sigma_\alpha,\sigma_\beta)}\right)^n, \tag{5.50}$$

where we defined:

$$f(\sigma_\alpha, \sigma_\beta) = \int \mathcal{D}z \, \log \left[H \left(\frac{\sigma_\alpha + \sigma_\beta - \sqrt{q^\alpha q^\beta} z}{\sqrt{1 - q^\alpha q^\beta}} \right) \right]. \tag{5.51}$$

It is immediate to derive the expression for the annealed free energy, *i.e.*:

$$\begin{aligned} F_A = -\frac{1}{\beta N} &\left[\log \int \prod_\alpha dW(\sigma_\alpha) \delta \left(\sum_\alpha \sigma_\alpha^p - \Sigma \right) e^{\sum_{\alpha \neq \beta} f(\sigma_\alpha, \sigma_\beta)} + \right. \\ &\left. + \frac{N}{2} \sum_\alpha \left(\log(1 - q^\alpha) + \frac{q^\alpha}{1 - q^\alpha} \right) \right] \end{aligned} \tag{5.52}$$

from which, in the absence of polydispersivity, *i.e.* if $W(\sigma_\alpha) = \delta(\sigma_\alpha - \Sigma/M)$, we recover back the quenched result of Eq. (5.48).

We may interpret the first term of the annealed free energy as the integral over the distribution of $\vec{\sigma}$ induced by the dynamics and define the probability distribution as:

$$P_q(\vec{\sigma}) \propto \delta \left(\sum_\alpha \sigma_\alpha^p - \Sigma \right) e^{\sum_{\alpha \neq \beta} f_q(\sigma_\alpha, \sigma_\beta)}, \tag{5.53}$$

where a normalization factor is implicitly embedded in the definition of $P_q(\vec{\sigma})$. To simplify the working scheme we might consider $q^\alpha = q^\beta = q$ and a uniform distribution for $W(\sigma_\alpha)$. In order to find the typical distribution of particles we notice that at a given q, we may run a Monte Carlo dynamics according to the following rules:

- pick at random two particles with associated variables σ_i and σ_j and allow the two-step process: $\sigma_i' = \sigma_i - \Delta$ and $\sigma_j' = \sigma_j + \Delta$, given the constraint $\sigma_i^p + \sigma_j^p = \sigma_i'^p + \sigma_j'^p$;

- compare the two energy values, *i.e.* $E = \sum_{\alpha \neq \beta} f_q(\sigma_\alpha, \sigma_\beta)$ and $E' = \sum_{\alpha \neq \beta}' f_q(\sigma_\alpha, \sigma_\beta)$ $+ 2 \sum_{\beta:\beta \neq i} f_q(\sigma_i', \sigma_\beta) + 2 \sum_{\beta:\beta \neq j} f_q(\sigma_j', \sigma_\beta)$, where we assumed that $f_q(\sigma_\alpha, \sigma_\beta) =$ $f_q(\sigma_\beta, \sigma_\alpha)$ and we denoted by $\sum_{\alpha \neq \beta}'$ the sum over all the indices except i and j;

- accept the move if and only if $E' > E$.

Starting from an initial $P(\vec{\sigma})$ at a given q, we wait for the system to equilibrate. We start collecting configurations only after a transient time to estimate the second term of the saddle point equation for q, which reads:

$$\frac{\alpha}{2}\frac{q^{\alpha}}{(1-q^{\alpha})^2} + \frac{1}{N^2}\sum_{\alpha\neq\beta}\langle\partial_q f_q(\sigma_\alpha,\sigma_\beta)\rangle_{P_q} = 0 \,. \tag{5.54}$$

Once a solution for this equation is known, we can update the value of the overlap q and iterate until convergence. From the resulting distribution of particle sizes, we can check whether the system undergoes a condensation transition or not.

Even without implementing a Monte Carlo simulation, we can draw some qualitative conclusions. The disclosed computation in high-dimensional sphere models leads us to answer the question about the emergence of a condensation transition by the negative. Let us try to better explain the underlying reasoning. To study the transition or a crossover regime, we should optmize the functional above with respect to $W(\vec{\sigma})$. However, looking at Eq. (5.52) for the free energy, we immediately understand that the terms depending on the actual size σ have a different order of magnitude: the energy term $\sum_{\alpha\neq\beta}f(\sigma_\alpha,\sigma_\beta)$ is proportional to $M(M-1)$, while the entropic term, which accounts for the symmetry under permutation of the particles,[3] is proportional to M only. Using the integral representation, the two contributions translate to:

$$-\beta FN \simeq M\int d\vec{\sigma}\,w(\vec{\sigma})\log(w(\vec{\sigma})) + \frac{M(M-1)}{2}\int d\vec{\sigma}d\vec{\sigma}'\,w(\vec{\sigma})w(\vec{\sigma}')f(\vec{\sigma},\vec{\sigma}') \,.$$
$$\tag{5.55}$$

The above expression is reminiscent of the analogous virial expansion usually resorted to in the liquid phase. However, the integration over $\vec{\sigma}, \vec{\sigma}'$ opens two different scenarios for the kernel $f(\vec{\sigma},\vec{\sigma}')$, depending on whether $f(\vec{\sigma},\vec{\sigma}')$ has a translational invariance and a fast-decay for $|\vec{\sigma}-\vec{\sigma}'|$ greater than a given length, or not. In the first scenario, the integral is $O(M)$ as the maximum length acts as a cut-off and crosses out the second integral. In the second scenario, without any translational invariance, $f(\vec{\sigma},\vec{\sigma}')$ is finite for any combination of particle sizes and in principle the integral should be $O(M^2)$. Note that two other terms, due to the normalization of the probability distribution and the constraint of the p-th moment, should be taken into account. We can nevertheless conclude that in this computation the entropic contribution is always sub-leading.[4] The leading behavior of the energetic term generates an exponential distribution for $W(\vec{\sigma})$ without any reason to justify the emergence of a condensate, unless a different suitable scaling of particle sizes and other control parameters is considered. Conversely, if the function $f(\sigma,\sigma')$ accounts only for finite-range or finite-connectivity interactions, the energetic term will be no longer proportional to the squared number of particles, but to the number of particles M times the average connectivity. In this regime, the two terms, namely the entropic contribution and the energetic one, can be comparable and induce a non-trivial particle distribution.

---

[3] The entropic term comes from the density $w(\sigma) = 1/M\sum_\alpha \delta(\sigma_\alpha - \sigma)$. Exponentiating the delta function and performing a saddle-point computation over the associated Lagrange multiplier, *i.e.* $\hat{w}(\sigma)$, we recover back a logarithmic contribution in $w(\sigma)$.

[4] This resembles the Coulomb gas with a temperature $T = 1/N$, where again the entropic term is subdominant.

5.4   Conclusions

We extended part of results of the previous chapter to derive a compact and well-defined free-energy functional for sphere systems. The small-coupling expansion can play a twofold role, allowing to better analyze the TAP equations, from an optmization and computer science perspective, but even to deepen the physics of jamming, from a condensed matter viewpoint. A detailed study of the effective potential in hard and soft-sphere systems is a central issue in the study of amorphous properties. The key outcome relies on the possibility of obtaining a suitable formalism without the need to introduce replicas nor virial expansions of the free energy.

From the analysis of high-dimensional hard-sphere models, what grabbed our attention was the possibility to properly describe the emergence of a sort of condensation transition[5] in polydisperse assemblies of hard spheres. This poses a non-trivial problem, already in low dimensions—except for $d = 1$—and it becomes more pronounced in the infinite-dimensional case. We addressed the problem by considering M spheres on a N-dimensional spherical manifold and then taking the thermodynamic limit for both degrees of freedom. Two different approaches have been pursued, one leading to the computation of the quenched free energy and another for the annealed case. In the first scenario we considered the radii of the particles as fixed and averaged over them. In the second scenario, the $\vec{\sigma}$'s are not *frozen* but they can contribute either in facilitating a phase transition or macroscopically modifying the structure of the system. This computation allowed us to distinguish the trivial (quenched) case from the annealed one, where nevertheless a condensation transition in real space looks unfeasible. The negative outcome is essentially linked to the fact that in high-dimensional models with a fully-connected topology most of the properties of real systems are missing. Despite the fact that such a transition seems to be absent in high dimension, it might be useful to exploit the considerations developed in this chapter to understand where a crossover can actually occur, what the most interesting regime is, etc., as the only analytical predictions are available just in very low dimension.

References

1. Abramowitz H, Segun I (1965) Handbook of mathematical functions. Dover, New York
2. Altieri A, Franz S, Parisi G (2016) The jamming transition in high dimension: an analytical study of the TAP equations and the effective thermodynamic potential. J Stat Mech: Theory Exp 2016(9):093301
3. Angel AG et al (2005) Critical phase in nonconserving zero-range processes and rewiring networks. Phys Rev E 72(4):046132
4. Bianconi G, Barabási A (2001) Bose-Einstein condensation in complex networks. Phys Rev Lett 86(24):5632
5. Bianconi G, Ferretti L, Franz S (2009) Non-neutral theory of biodiversity. EPL (Europhys Lett) 87(2):28001

---

[5]This terminology is meant for purely classical systems.

6. Bianconi G et al (2011) Modeling microevolution in a changing environment: the evolving quasispecies and the diluted champion process. J Stat Mech: Theory Exp 2011(08):08022

7. Biroli G, Mézard M (2001) Lattice glass models. Phys Rev Lett 88(2):025501

8. Bouchaud J-P, Mézard M (1994) Self induced quenched disorder: a model for the glass transition. Journal de Physique I 4(8):1109

9. Bouchaud J-P, Mézard M (2000) Wealth condensation in a simple model of economy. Phys A: Stat Mech Appl 282(3–4):536

10. Ciamarra MP et al (2003) Lattice glass model with no tendency to crystallize. Phys Rev E 67(5):057105

11. Cugliandolo LF et al (1996) A mean-field hard-spheres model of glass. J Phys A: Math Gen 29(7):1347

12. Evans MR et al (2010) Condensation transition in polydisperse hard rods. J Chem Phys 132(1):014102

13. Frisch HL, Percus JK (1999) High dimensionality as an organizing device for classical fluids. Phys Rev E 60(3):2942

14. Hansen IR, McDonald J-P (1990) Theory of simple liquids. Elsevier

15. Kurchan J, Maimbourg T, Zamponi F (2016) Statics and dynamics of infinitedimensional liquids and glasses: a parallel and compact derivation. J Stat Mech: Theory Exp 2016(3):033210

16. Kurchan J, Parisi G, Zamponi F (2012) Exact theory of dense amorphous hard spheres in high dimension. I. The free energy. J Stat Mech: Theory Exp 2012(10):10012

17. Mari R, Krzakala F, Kurchan J (2009) Jamming versus glass transitions. Phys Rev Lett 103(2):025701

18. Mari R, Kurchan J (2011) Dynamical transition of glasses: from exact to approximate. J Chem Phys 135:124504

19. Marinari E, Parisi G, Ritort F (1994) Replica field theory for deterministic models: I. Binary sequences with low autocorrelation. J Phys A: Math Gen 27(23):7615

20. Maynar P, Trizac E (2011) Entropy of continuous mixtures and the measure problem. Phys Rev Lett 106(16):160603

21. O'Loan OJ, Evans MR, Cates ME (1998) Jamming transition in a omogeneous one-dimensional system: the bus route model. Phys Rev E 58(2):1404

22. Parisi G, Slanina F (2000) Toy model for the mean-field theory of hard-sphere liquids. Phys Rev E 62(5):6554

23. Parisi G, Zamponi F (2010) Mean-field theory of hard sphere glasses and jamming. Rev Mod Phys 82(1):789

24. Percus JK, Yevick GJ (1958) Analysis of classical statistical mechanics by means of collective coordinates. Phys Rev 110(1):1

25. Rivoire O et al (2004) Glass models on Bethe lattices. Eur Phys J B 37:5

26. Salacuse JJ, Stell G (1982) Polydisperse systems: statistical thermodynamics, with applications to several models including hard and permeable spheres. J Chem Phys 77(7):3714

27. Török J (2005) Analytic study of clustering in shaken granular material using zero-range processes. Phys A: Stat Mech Appl 355(2–4):374

28. Zhang J et al (1999) Optimal packing of polydisperse hard-sphere fluids. J Chem Phys 110(11):5318

Chapter 6
The Jamming Paradigm in Ecology

As discussed in the previous chapters, in the very last years the physics of jamming has gained momentum in several interdisciplinary contexts, ranging from machine learning to inference, ecology and beyond. Nowadays, jamming appears to be a wide-ranging phenomenon, which can be recovered in fields apparently disconnected from structural glasses, but in practice more related than one might think at first sight. One of the most innovative and challenging aspects of this chapter is the definition of complete statistical mechanics models that allow us to describe and fully capture the deep network of connections between rough landscapes in glassy systems and ecosystems at the edge of stability. This analysis has been published recently in [1].

6.1 Stability and Complexity in Ecosystems

Critical properties of complex systems have been the object of an intense study in the last decades, both theoretically and experimentally. On a theoretical level, such systems of single interacting agents are extremely common and range from economy to biophysics, social science, and game theory as well. Motivated by a renewed interest in ecology, due in particular to numerous experimental results and more advanced techniques, in the last few years physicists have attempted to describe emergent, collective phenomena in living systems using tools and intuitions from statistical mechanics. On the one hand, a great revolution in microbial ecology, thanks also to the employment of engineered bacteria, has allowed to unveil the complexity of microorganism communities. On the other hand, classical models of resource-competition can be formulated as a problem of statistical physics of disordered systems and then solved analytically in the large-size limit, by taking advantage of the definition of *self-averaging* quantities [6]. Taking each detail and each individual behavior into account can go beyond any reasonable human effort.

© Springer Nature Switzerland AG 2019
A. Altieri, *Jamming and Glass Transitions*, Springer Theses,
https://doi.org/10.1007/978-3-030-23600-7_6

However, as long as one is concerned with collective properties, a reasonable way to address the problem is to model the system via random couplings [20].

In the following, we shall mostly focus on this kind of problems, with a special emphasis on equilibrium properties and stability.[1]

The interesting mechanisms existing between organisms and their environment have been largely studied both at equilibrium, i.e. in resource-competition models, and out of equilibrium, i.e. in ecological succession problems.[2] Generally, for population models in a deterministic environment, one looks for community equilibria where all populations eventually belong to different species and reach constant values. According to this picture, a system can be stable or unstable depending on its response to external perturbations. If the populations tend to their equilibrium value, despite the application of an external perturbation, the system is said to be stable; otherwise, if the perturbation gets amplified in time, the system is unstable. Both behaviors can be characterized by damped (or undamped) oscillations (as qualitatively shown in Fig. 6.1) or by a monotonic trend. In the case of a community of p species, the simplest way to proceed is to represent the solution of the population-dynamics equations on a certain p-dimensional surface. A point on this surface corresponds to a specific set of populations. The emergence of an equilibrium condition is then identified by a flat landscape. Conversely, the presence of valleys or hilltops is a condition for stability and instability, respectively. Another important difference concerns the linearity or non-linearity of the dynamical differential equations governing ecological or biological populations. Only in the first case, neighborhood stability and global stability coincide. In general, these two notions are different and lead to a complex non-uniform landscape. Moreover, the non-linearity can further complicate the analysis and generate limit cycles which make the populations change cyclically in time. A paradigmatic example is given by the Lotka-Volterra model [10, 19], in which sustained oscillations and limit cycles are somehow pathological. In principle, to characterize the multiplicity and complexity of scenarios occurring in an ecosystem, one can try to define a Lyapunov function. The definition of such a function provides a mathematically precise way to characterize the stability of a general competing system against perturbations. However, in several specific cases, as for instance in the presence of asymmetric interactions, a more involved approach is needed.

Let us suppose to consider a system described by the following differential equation:

$$\frac{dN}{dt} = F(N) , \tag{6.1}$$

---

[1] It is worth noticing, however, that while equilibrium and homogeneity are central concepts in physics, non-equilibrium and heterogeneity are often the driving mechanisms in ecology and finance as well.

[2] In this context, the concepts of *equilibrium* and *steady-state* can be confused or wrongly employed. A stationary state to remain constant requires a continual work (via energy or mass injection), differently than a system in equilibrium, where there is no incoming or leaving free energy. Clearly, in the absence of dissipation and external renewing sources, detailed balance would be preserved and the concepts of *equilibrium* and *stationary state* would coincide.

Fig. 6.1 Stylization of the typical behavior of a stable system as a function of time describing the solution of a generic differential equation. Both curves represent damped oscillations against an external perturbation. The second curve in yellow is obtained in the presence of small noise

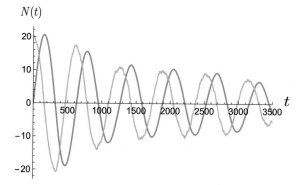

where N denotes the population evolving in time. $F : \Omega \subset R^n \to R^n$ is a continuous smooth function satisfying the condition $F(N^*) = 0$. We indicate with U the neighborhood of the equilibrium point N^*. A scalar function $\mathcal{L} : U \to \mathbb{R}$ is called *Lyapunov function* [11] if it is continuously differentiable, positive-definite, i.e.:

$$\begin{cases} \mathcal{L}(N) > 0 & \text{if} \quad N \neq N^* \\ \mathcal{L}(N) = 0 & \text{if} \quad N = N^* \end{cases} \tag{6.2}$$

and additionally if:

$$\nabla \mathcal{L}(N) \cdot \frac{dN}{dt} \leq 0 . \tag{6.3}$$

This relation guarantees the stability of the zero solution in Eq. (6.1).[3] In other words, the sign of the Lyapunov function and the last condition allow to understand whether the equilibrium is stable, locally asymptotically stable or globally asymptotically stable. One can then study the system behavior upon perturbing it from N^* within its stability domain. If the attraction domain is sufficiently large, the system is more able to sustain perturbations. Similarly, the steeper the Lyapunov function—as denoted by $\nabla \mathcal{L}(N)$—the faster the system moves to its initial value N^*. The first property is usually called *tolerance*, while the second one is known as *resilience*.

To generalize this argument we can consider a multi-species community formed by p populations, which are labeled by index i. Its dynamics is defined by:

$$\frac{dN_i(t)}{dt} = F_i(N_1, (t), ..., N_p(t)) , \tag{6.4}$$

where F_i is a non-linear function of the interacting populations. The equilibrium condition requires that $F_i(N_1^*, ..., N_p^*) = 0$. We can then slightly perturb the solution

---

[3]The first Lyapunov function in population dynamics was introduced by Volterra in the twenties [3], considering a stylised prey-predator model for describing mutual fluctuations of sharks and food fish in the Mediterranean sea.

such that:

$$N_i(t) = N_i^* + x_i(t) . \tag{6.5}$$

The perturbation $x_i(t)$ satisfies:

$$\frac{dx_i(t)}{dt} = \sum_{j=1}^{m} a_{ij} x_j(t) , \tag{6.6}$$

which has been obtained by Taylor expanding Eq. (6.4) around the equilibrium point and neglecting second-order terms. The elements a_{ij} are defined as:

$$a_{ij} = \left(\frac{\partial F_i}{\partial N_j} \right) \bigg|_{N^*} . \tag{6.7}$$

They represent the so-called *community matrix* elements [14] and describe the effect of the species j upon the species i near equilibrium. By solving Eq. (6.6), one obtains:

$$x_i(t) = \sum_{j=1}^{m} C_{ij} \exp(\lambda_j t) , \tag{6.8}$$

where the coefficients C_{ij} depend on the initial values of the external perturbation, while the λ_j's are the eigenvalues of the community matrix. From the analysis of the eigenvalues, one can obtain information on the stability, mostly determined by the real part of the eigenvalues. For instance, if one or more eigenvalues are purely imaginary, while the others have negative real parts, the stability is said to be *neutral*. The presence of a single eigenvalue with a positive real part is sufficient to destabilize the whole system. As we mentioned before, in a multispecies community, there is a great difference between global stability and neighborhood stability. However, if all eigenvalues of the community matrix lie in the left-hand side of the complex plane, the neighborhood stability is assured.

As a matter of fact, the elements of the community matrix contain much more information of direct biological relevance. Given the pair of matrix elements a_{ij} and a_{ji}, one can distinguish five different interaction types depending on the mutual signs of the elements: (i) +0 stands for *commensalism*; (ii) −0 for amensalism; (iii) ++ represents instead a *symbiosis* interaction; (iv) −− stands for a *competion*; (v) +− includes a *prey-predator* effect. This sign structure starting from the community matrix analysis was proposed first by Odum [15]. This kind of reasoning is nevertheless feasible in low dimensions only. In the presence of large space dimensions or of an increasing number of species, it can become truly hard and counterintuitive.

In the past years, however, there has certainly been a considerable interest in theoretical ecology. Different models have been proposed taking into account, for instance, either discrete or continuous growth, stochasticity in the demographic evolution as well as in the environment, symmetric or antisymmetric interactions, etc.

One might ask whether changing the connectivity, the degree of heterogeneity or the strength of the interactions can affect the stability of the system. More interestingly, recent works proposed criticality and glassiness as emergent properties of ecosystems [7, 9].

Another worth noticing aspect is the distinction between a thermodynamic/static approach and a dynamical analysis. In principle, the latter gives access to a broader spectrum of questions. Indeed, while static approaches are limited to the existence of a Lyapunov function and then to symmetric interactions only among individuals, a dynamical formalism allows to extend the analysis to cases with asymmetric interaction matrices. In the following, we shall focus on a static formalism specifically applied to a variant of the original MacArthur's model [12], as recently proposed by Tikhonov and Monasson [16, 17]. MacArthur's model has the advantage to be one of the simplest models and also one of the first—dating back to the end of the sixties—that has been used to analyze ecological systems in terms of stability and complexity.

6.2 MacArthur's Model

This model was originally introduced to describe resource-competition systems, for which Tilman proposed a geometrical interpretation with few species and resources [18]. Resources are considered non-interacting variables and any competition among species is only due to the exploitation of the available resources.

An interesting aspect relies on its high-dimensional generalization because of the emergence of intriguing and complex behaviors, which are actually missing in dimensions one or two. The environment is intrinsically a high-dimensional object, described in terms of statistical mechanics of many-body systems. However, most of the predictions and results that can be immediately derived in low dimension do not apply to high dimension. The consequence of this failing approach can be understood in MacArthur's model by simply increasing the number of resources to ten or fifteen units. In its original formulation [12], the ecosystem environment is described in terms of the abundances n_μ of the species μ, which satisfy the relation:

$$\frac{dn_\mu}{dt} = c_\mu n_\mu \left(\sum_i w_i \sigma_{i\mu} h_i - \chi_\mu \right) . \tag{6.9}$$

The index $\mu = 1, ..., S$ labels the number of the species up to S, h_i stands for the abundance of resource i ($i = 1, .., N$), with an associated weight w_i, while $\sigma_{i\mu}$ is the probability per unit time that an individual of the species μ exploits a certain unit of the resource i. Below a given threshold χ_μ, the species μ start decreasing. An analogous differential equation can be written for the resource growth:

$$\frac{dh_i}{dt} = \frac{r_i h_i}{K_i} (K_i - h_i) - \sum_\nu \sigma_{i\nu} n_\nu h_i \,, \tag{6.10}$$

where we indicate as K_i the carrying capacity of the resource h_i in the environment and as r_i the *intrinsic rate of natural increase*. Eqs. (6.9)–(6.10) are similar to the Lotka-Volterra equations with a slight modification due to the second term in Eq. (6.10) acting as a self-limitation. The condition for equilibrium is then:

$$h_i = K_i - \sum_\nu \frac{K_i}{r_i} \sigma_{\nu i} n_\nu = K_i \left(1 - \frac{T_i}{r_i} \right) \,, \tag{6.11}$$

where T_i is the total demand. Therefore, from Eq. (6.11) we can derive the following relation, which describes the evolution of the species abundances:

$$\frac{dn_\mu}{dt} = c_\mu n_\mu \left[\left(\sum_i w_i \sigma_{i\mu} K_i - \chi_\mu \right) - \sum_\nu \left(\sum_i \sigma_{i\nu} \sigma_{i\mu} \frac{w_i K_i}{r_i} \right) n_\nu \right] \,. \tag{6.12}$$

At equilibrium the variation of n_μ must be zero, forcing the two terms in parentheses to balance each other out. This condition can be also obtained by minimizing a suitable quadratic function with respect to the abundances, i.e.:

$$Q = \sum_i \frac{w_i K_i}{r_i} (r_i - n_1 \sigma_{1i} - n_2 \sigma_{2i} - \ldots)^2 + 2 \sum_\nu \chi_\nu n_\nu \,. \tag{6.13}$$

This equation basically guarantees the square of unused production to be as small as possible if a contribution $\sum_\nu \chi_\nu n_\nu$ is added to the food requirement. Only positive values of that expression should be taken into account because, if any of the terms in parenthesis is negative, there will be no contribution from the corresponding resource i. In the original MacArthur model, χ_μ is assumed to be constant as well as $\sum_i \sigma_{\nu i}$, meaning that all species have the same total ability to harvest nourishment. Then, the equilibrium condition provided by Eq. (6.12) is equivalent to minimizing something even simpler.

6.2.1   Beyond the MacArthur Model: The Role of High Dimensionality

MacArthur's model has been used as a platform for the following analysis, whose generalization to large space dimension turns out to be more suitable to be treated in the framework of statistical physics of disordered systems. We shall first introduce the Tikhonov-Monasson model, which is exactly solvable in the thermodynamic limit.

Then we shall present our recent results concerning the analysis of the spectrum of small fluctuations of the Lyapunov function defined for this specific model.

According to [17], a well-mixed habitat[4] is considered where any element i exists in N different forms. These N forms represent the available resources for a population of S different species. For simplicity, we will use the same notations both for MacArthur's model and for its generalized version.

The number of individuals n_μ, where again μ denotes the different species, depends on the availability of resources, h_i, which are a function the total demand $T_i = \sum_\mu n_\mu \sigma_{\mu i}$:

$$h_i = \frac{R_i}{\sum_\mu n_\mu \sigma_{\mu i}} , \qquad (6.14)$$

R_i is the resource supply whose average value is assumed to be constant, while its variance represents a control parameter of the model and describes heterogeneity in the resources (for more details, see the phase diagram in Fig. 6.2). The matrix $\sigma_{\mu i}$ is called *metabolic strategy* through which a species demands and meets its requirement χ_μ. For each species μ, the vector $\sigma_{\mu i}$ is chosen randomly and takes values 0 and 1 with probabilities $(1 - p)$ and p respectively.[5] Compared to Eq. (6.11) the availabilities are chosen here as a decreasing function of the total demand, T_i.

Then, the control parameters of the model can be summarized in: $\alpha = S/N$, the *density* of the species pool; ϵ, the width of the cost distribution; p, the average of the metabolic strategy distribution and $\overline{\delta R^2}$, the variance in the resource supply. The dynamics of the model is thus defined by the following differential equation:

$$\frac{dn_\mu}{dt} \propto n_\mu \Delta_\mu . \qquad (6.15)$$

and, because we are interested in the equilibrium condition, any proportionality factor can be safely neglected. The quantity Δ_μ represents the *resource surplus*, namely:

$$\Delta_\mu = \sum_{i=1}^{N} \sigma_{\mu i} h_i - \chi_\mu . \qquad (6.16)$$

where χ_μ is the species requirement. Comparing Eqs. (6.15) with (6.12) one immediately sees that only the first term in parenthesis contributes here, leading to a further simplified version.

---

[4]In a well-mixed habitat each agent interacts with all other agents: there is no notion of distance, in line with the instances of fully-connected systems described so far.

[5]The parameter p has a deeper physical meaning: it is actually responsible for making the phase transition, which will be shown in the phase-diagram below, in Fig. 6.2, more or less pronounced. It is also important for distinguishing between a specialist and a generalist in the typical competitor classification: it determines whether the species in the ecosystem are either specialists ($p \ll 1$), each requiring a small amount of well-defined metabolites, or generalists ($p \sim 1$), interested in many different metabolites.

One can also re-define the resource surplus by a random cost, i.e. $\chi_\mu = \sum_i \sigma_{\mu i} + \epsilon x_\mu$, where ϵ is a formal infinitesimal parameter eventually set to zero, while x_μ is a Gaussian variable with unit variance. The formalism we are going to employ in the following lines is built on the—quite strong—assumption that the strategy and its cost are uncorrelated variables. Then, the resource surplus becomes:

$$\Delta_\mu = \sum_i \sigma_{\mu i} (h_i - 1) - \epsilon x_\mu \,. \tag{6.17}$$

Within this change of variables, the fluctuations of the model are a function of $(h_i - 1)$, as well as the other physical quantities of the model, which can be reinterpreted in terms of:

$$m = \sum_i (1 - h_i) \,, \qquad q = \sum_i (1 - h_i)^2 \,. \tag{6.18}$$

The average value of the resource surplus over μ is:

$$\langle \Delta_\mu \rangle = \left\langle \sum_i (h_i - 1) \sigma_{\mu i} - \epsilon x_\mu \right\rangle = -pm \,, \tag{6.19}$$

while $\psi = \sqrt{p(1 - p)q + \epsilon^2}$ represents its spread. We will make heavy use of these two parameters in the following to detail our stability analysis. To perform our computation we first need to consider the equilibrium condition:

$$\frac{dn_\mu}{dt} = 0 \,, \tag{6.20}$$

which leads to two different possibilities: (i) $n_\mu > 0$ & $\Delta_\mu = 0$; (ii) $n_\mu = 0$ & $\Delta_\mu < 0$. To be more precise, a stable equilibrium, with no invasion, is characterized by an additional condition, namely that all species with $n_\mu = 0$ tend to become extinct. In other words, the equilibrium state is detectable once the solution of the corresponding convex optimization problem on the boundary phase is known. This model can be indeed interpreted as a constraint satisfaction problem (CSP): from this perspective, a positive resource surplus allows species to survive and multiply; conversely, if $\vec{h} \cdot \sigma_\mu < \chi_\mu$, resource availability cannot guarantee the sustainability of the species' pool. All $\vec{h}$ such that $\vec{h} \cdot \sigma_\mu < \chi_\mu$ define the so-called *unsustainable region*.

In [17] it has been proven that this model undergoes a phase transition between two qualitatively different regimes. Focusing on the phase diagram in the left panel of Fig. 6.2, one can distinguish a *shielded phase S*, wherein the internal environment acquires a collective behavior due only to the species and independent of external conditions. The feedback mechanism contributes to adjusting mutual species' abundance keeping the availabilities to one, for all i. The situation is quite different in the *vulnerable phase* phase V where the species cannot self-sustain, strongly affected by changes and improvements in the immediate environment. In our feeling, the S phase should correspond to an isostatic phase, while the V phase should lead to a hypostatic regime. We used here the terminology borrowed from the jamming con-

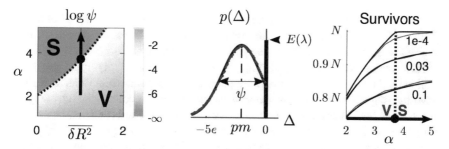

Fig. 6.2 In the left panel we report the phase diagram of the model in the $\epsilon \to 0$ limit. The critical dotted line separates a S phase from a V phase. In the center, the distribution of resource surplus at equilibrium is plotted: the theoretical prediction is in black and the simulated behavior over 500 realizations at $N = 50$ in red. In the rightmost panel the qualitative trend of the number of surviving species as a function of α is shown. Figures reprinted from [17]

text. Originally, in [17] the authors did not focus on this concept, which grabbed our attention and gave the true motivation to the following analysis. Our terminology is inspired by the fact that, upon increasing α, the system enters the S phase, where the number of *active* species equals the number of resources, basically the same picture that has been recovered in Chap. 4 while discussing the perceptron behavior at jamming.

6.3 Similarities with the Percetron Model

To highlight the analogies between the disclosed ecological model and the perceptron in the convex phase, we can start introducing the distribution of the resource surplus at equilibrium:

$$p(\Delta) = \frac{1}{\sqrt{2\pi\psi^2}} e^{-\frac{(\Delta+pm)^2}{2\psi^2}} \theta(-\Delta) + H\left(\frac{pm}{\psi}\right)\delta(\Delta) , \qquad (6.21)$$

where the δ function contribution accounts for the fraction of species that actually reaches its food requirement. According to our notation, $H(x) \equiv \int_x^\infty \mathcal{D}z$. The condition $1 - \alpha H((pm)/a) = 0$ properly identifies the *shielded* phase, providing a vanishing value inside that and finite otherwise.

Similarly, the distribution of gaps $g(h)$ in the perceptron near jamming- not to be confused with the availability h_i—always in the RS phase, can be written as [8]:

$$g(h) = \alpha(1 - H(\sigma))\delta(h) + \frac{\alpha}{\sqrt{2\pi}} e^{-\frac{(h+\sigma)^2}{2}} \theta(h) . \qquad (6.22)$$

As explained in Chap. 4, the transition from a hypostatic to an isostatic region is determined by the control parameter σ. For $\sigma > 0$ the system is hypostatic and the coefficient of the delta-function gives the fraction of binding constraints. For $\sigma < 0$, the space of solutions becomes disconnected and each constraint defines a non-convex domain. We recall from the perceptron model that Eq. (6.22) is no longer valid in the non-convex regime leading to a power-law singularity in the gap distribution for small values of h. This phase precisely corresponds to an isostatic regime. Investigating whether a similar scenario can also occur in ecology and unveiling the mechanism behind it might be an interesting research topic and open new fascinating directions. In the following, we briefly recall some outcomes of [17], first to highlight the links with the convex perceptron near the jamming onset, and then to introduce our new results, which will be extensively presented in Sects. 6.4 and 6.5.

The partition function of the model recently proposed by Tikhonov and Monasson can be written as [17]:

$$Z = \int_0^\infty \prod_\mu dn_\mu \int_0^\infty dT_i \delta(T_i - \sum_\mu n_\mu \sigma_{\mu i}) e^{\beta F(n_\mu)} , \qquad (6.23)$$

where $T_i = \sum_\mu n_\mu \sigma_{\mu i}$ denotes the total demand. By introducing the exponential representation of the δ-function, Eq. (6.23) becomes:

$$Z = \int_0^\infty \prod_\mu dn_\mu \int_0^\infty dT_i \left[\int \frac{d\theta_i}{2\pi/\beta} e^{-i\beta\theta_i(T_i - \sum_\mu n_\mu \sigma_{\mu i})} \right] e^{\beta\left(\sum_i \hat{H}(T_i) - \sum_\mu n_\mu \chi_\mu\right)} ,$$

$$(6.24)$$

where $\hat{H}(x) = \int^x H(T) dT$. At this level, one can integrate over n_μ and obtain:

$$Z = \int_0^\infty dT_i \int \frac{d\theta_i}{2\pi/\beta} e^{\beta \sum_i \left[\hat{H}(T_i) - i\theta_i T_i\right]} \prod_\mu \frac{1/\beta}{\chi_\mu - i \sum_i \theta_i \sigma_{\mu i}} . \qquad (6.25)$$

For large β, the integral over T_i can also be evaluated using the saddle-point approximation, which implies that $H(T_i^*) = i\theta_i \equiv h_i$. This makes the introduction of the Legendre transform more direct:

$$Z = const. \cdot \int_{-i\infty}^{+i\infty} d\vec{h} e^{-\beta \tilde{F}(\vec{h})} \prod_\mu \frac{1/\beta}{\chi_\mu - \vec{h} \cdot \vec{\sigma}_\mu} . \qquad (6.26)$$

For large β this problem turns again into a saddle-point evaluation. However, the integration contour cannot simply be rotated onto the real axis because of the complicated pole structure of the integrating function. One need to deform the integration contour and to evaluate it over piecewise-linear domains. However, the integral can be computed only if the poles are avoided and all the availabilities have real values. Then, the most interesting regime, which is delimited by the hyperplanes that satisfy these constraints for the availabilities, corresponds to the *unsustainable region* Ω, where Δ_μ is negative. The case $\Delta_\mu = 0$ is excluded because it corresponds to a pole in the integrand. Therefore, one has:

$$Z = const. \cdot \int_{\Omega} d\vec{h} e^{-\beta \tilde{F}(\vec{h})} \prod_{\mu} \frac{1/\beta}{\chi_{\mu} - \vec{h} \cdot \vec{\sigma}_{\mu}} . \tag{6.27}$$

where the function $\tilde{F}$ is the Legendre transform of $\hat{H}(x) = \int^x H(T) dT$, namely:

$$\tilde{F}(\vec{h}) = \left(\vec{h} \cdot \vec{T} - F(T) \right)\Big|_{T^*} \tag{6.28}$$

$$\tilde{F} = \sum_i h_i \left(\sum_{\mu} n_{\mu} \sigma_{\mu i} \right) - \sum_i \hat{H}(T_i) . \tag{6.29}$$

In Eq. (6.27) the exponential factor dominates everywhere except in the region where $\frac{1}{\Delta_{\mu}}$ diverges, i.e. near the boundary. The key point lies in the difference between the finite-temperature regime and the zero-temperature limit: indeed, in the former all species n_{μ} are characterized by a finite abundance,[6] while in the latter only a subset does not vanish, which corresponds to the species with Δ_{μ} exactly zero. In this second case the extremum of the integral lies on the boundary of Ω.

Another interesting result of this models concerns the emergence of a logarithmic trend in Δ_{μ} at large but finite β:

$$\log Z = \max \left[-\beta \tilde{F}(\vec{h}) - \sum_{\mu} \log |\Delta_{\mu}| \right] . \tag{6.30}$$

by computing the logarithm of the partition function in Eq. (6.27). The logarithmic contribution becomes dominant in the low-temperature regime when Δ_{μ} tends to zero for a few selected species. Equation (6.30) is reminiscent of an analogous result already discussed for the perceptron model in Chap. 4 upon approaching the jamming line from the SAT phase. In that case, the analogous logarithmic behavior was a consequence of a marginal mechanical stability, as a function of the random gaps h_{μ} between particles, which are somehow related to the resource surplus here.

6.4 Spectral Density of Harmonic Fluctuations of the Lyapunov Function

This model can be described in terms of a simple Lyapunov function that can be also proven to be positive-definite and convex. It is basically the building block to develop our stability analysis. The Lyapunov function is convex and bounded from above, which guarantees that an equilibrium state always exists.

---

[6]In this first case the distribution of the abundances follows an exponential law with average value $\langle n_{\mu} \rangle = 1/ \left(\beta |\Delta_{\mu}| \right)$.

$$\tilde{F} = \sum_i h_i \left(\sum_\mu n_\mu \sigma_{i\mu} \right) - \sum_i \hat{H}(T_i) , \qquad (6.31)$$

whose first derivative with respect to the availabilities h_i yields:

$$\frac{d\tilde{F}}{dh_i} = \frac{\partial \tilde{F}}{\partial h_i} + \frac{\partial \tilde{F}}{\partial T_i} \frac{\partial T_i}{\partial h_i} - \frac{\partial T_i}{\partial h_i} \frac{\partial \hat{H}(T_i)}{\partial T_i} , \qquad (6.32)$$

which can further be simplified thanks to the condition $H(T_i) = h_i$. To evaluate the second derivative of $\tilde{F}$, we also need to know $\frac{\partial n_\mu}{\partial h_j}$, that is:

$$\frac{\partial n_\mu}{\partial h_j} = \frac{\partial n_\mu}{\partial \Delta_\nu} \frac{\partial \Delta_\nu}{\partial h_j} , \qquad (6.33)$$

where we assumed to sum over repeated indices. Therefore, recalling the definition of the resource surplus Δ_μ in Eq. (6.16), we immediately end up with:

$$\frac{\partial \Delta_\mu}{\partial n_\nu} = - \sum_i \sigma_{\mu i} \sigma_{\nu i} \frac{R_i}{(\sum_\rho \sigma_{\rho i} n_\rho)^2} = - \sum_i \sigma_{\mu i} \sigma_{\nu i} \frac{h_i^2}{R_i} . \qquad (6.34)$$

Tethering together all the information thus far, the expression for the stability matrix $\frac{d^2 \tilde{F}}{dh_i dh_j}$, as a function of the availabilities, can be eventually obtained.

An alternative and more promising approach might be to consider the dual problem, by defining the stability matrix in terms of the species abundances n_μ. We can thus consider the following definition of the Lyapunov function [17]:

$$F(\{n_\mu\}) = \sum_i \hat{H}(T_i) - \sum_\mu n_\mu \chi_\mu = \sum_i R_i \log \left(\sum_\mu n_\mu \sigma_{\mu i} \right) - \sum_\mu n_\mu \chi_\mu , \quad (6.35)$$

to be differentiated with respect to the abundances:

$$\frac{dF}{dn_\mu} = \sum_i H(T_i) \frac{\partial T_i}{\partial n_\mu} - \chi_\mu = \Delta_\mu . \qquad (6.36)$$

Taking the second derivative on it, we get:

$$\frac{d^2 F}{dn_\mu dn_\nu} = - \sum_i \sigma_{\mu i} \sigma_{\nu i} \frac{R_i}{(\sum_\rho n_\rho \sigma_{\rho i})^2} = - \sum_i \sigma_{\mu i} \sigma_{\nu i} \left(\frac{h_i^2}{R_i} \right) . \qquad (6.37)$$

In the simplest case, for $h_i = 1$, the stability matrix reduces to a Wishart matrix whose distribution of eigenvalues is described by a Marchenko-Pastur law [13] in the large-N limit. Borrowing the results presented for the perceptron model, we can

Fig. 6.3 Qualitative behavior of the spectral density $\rho(\lambda)$ for a generic Marchenko-Pastur distribution (full black) with arbitrary λ_- and λ_+. The corresponding distribution with $\lambda_- = 0$ is plotted in dashed grey. The second one gives the leading contribution of the spectral density in the *shielded phase*

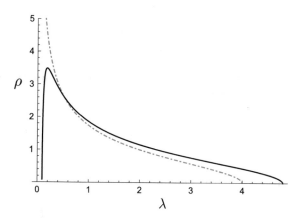

rewrite here:

$$\rho(\lambda) = \frac{1}{2\pi} \frac{\sqrt{(\lambda - \lambda_-)(\lambda_+ - \lambda)}}{\lambda} , \tag{6.38}$$

(see Fig. 6.3 above), where the upper and lower edges are $\lambda_\pm = (\sqrt{[1]} \pm 1)^2$, since no diagonal term appears in Eq. (6.37).[7] Then the leading contribution in the S phase turns out to be [1]:

$$\rho(\lambda) \sim \sqrt{\frac{4 - \lambda}{\lambda}} . \tag{6.39}$$

By contrast, in the most general case, with $h_i \neq 1$, we need to evaluate the full distribution of the availabilities h_i in Eq. (6.37) and their associated moments. Computing the full expression of the availability distribution is a possibility. Another possibility is to evaluate the eigenmodes of the stability matrix and their asymptotic behavior in the different phases. This analysis would allow us to justify the S/V transition in the $\epsilon \to 0$ limit and to recover a clear signature of the marginally stable nature of the system.

6.5 Stability Analysis: The Replicon Mode

In order to compute the replicon eigenvalue, which is the dominant eigenvalue of the stability matrix at criticality, we consider the Legendre transform of Eq. (6.23). By defining $h_i \equiv 1 - g_i/N$, one can write the partition function as [17]:

---

[7]The quantity [1] multiplied by N provides the number of effective interactions at criticality. This notation allows us to highlight once more the connection with the perceptron.

$$Z = \int_{-\infty}^{N} \prod_{i=1}^{N} \frac{dg_i}{N} e^{\beta \tilde{F}(\{g_i\})} \prod_{\mu=1}^{P} \int d\Delta_\mu \theta(-\Delta_\mu)\delta\left(\Delta_\mu + \epsilon x_\mu + \frac{1}{N} \sum_i g_i \sigma_{\mu i}\right)$$

$$(6.40)$$

and introduce the following quantities to make use of the replica method and consequently decouple indices μ and i:

$$m^a \equiv \frac{1}{N} \sum_i^N g_i^a, \qquad Q^{ab} \equiv \frac{1}{N^2} \sum_i^N g_i^a g_i^b .$$

$$(6.41)$$

The replicated partition function eventually reads [17]:

$$\overline{Z^n} = \int \prod_{a \leq b} \frac{dQ^{ab} d\hat{Q}^{ab}}{2\pi} \int \prod_a \frac{dm^a d\hat{m}^a}{2\pi} \exp\left[i \sum_{a \leq b} Q^{ab} \hat{Q}^{ab} + i \sum_a \hat{m}^a m^a\right] \cdot$$

$$\prod_i \left\{ \int \prod_a \frac{dg_i^a}{N} \exp\left[\sum_a \beta \tilde{F}_i(g_i^a) - \frac{i}{N} \sum_a \hat{m}^a g_i^a - \frac{i}{N^2} \sum_{a \leq b} \hat{Q}^{ab} g_i^a g_i^b\right]\right\} \cdot$$

$$\prod_\mu \left\{ \int \prod_a \frac{d\Delta_\mu^a d\hat{\Delta}_\mu^a}{2\pi} \prod_a \theta(-\Delta_\mu^a) \exp\left[i \sum_a \hat{\Delta}_\mu^a(\Delta_\mu^a + pm^a) +\right.\right.$$

$$\left.\left. - \frac{1}{2} \sum_{a,b} \left(p(1-p)Q^{ab} + \epsilon^2\right) \hat{\Delta}_\mu^a \hat{\Delta}_\mu^b\right]\right\} .$$

$$(6.42)$$

This definition of the replicated partition function is borrowed from [17]. On top of that, we can go into more detail and perform our stability analysis. In the following we will indicate the expression at the exponent as $\mathcal{S}$ on which we will perform a saddle-point computation, namely:

$$\mathcal{S} \propto \left[i \sum_{a \leq b} Q^{ab} \hat{Q}^{ab} + i \sum_a \hat{m}^a m^a - \frac{i}{N} \sum_i \sum_a \hat{m}^a g_i^a - \frac{i}{N^2} \sum_i \sum_{a \leq b} \hat{Q}^{ab} g_i^a g_i^b +\right.$$

$$\left. + i \sum_\mu \sum_a \hat{\Delta}_\mu^a(\Delta_\mu^a + pm^a) - \frac{1}{2} \sum_\mu \sum_{a,b} \left(p(1-p)Q^{ab} + \epsilon^2\right) \hat{\Delta}_\mu^a \hat{\Delta}_\mu^b\right] .$$

$$(6.43)$$

Once the second derivatives with respect to Q^{ab}, $\hat{Q}^{ab}$, m^a, $\hat{m}^a$ and so forth are known, we will be able to reconstruct the full expression of the Hessian matrix. It bears four replica indices and in principle its diagonalization in real space can be a hard task. We can nevertheless approach the transition line from above, i.e. from the so-called paramagnetic phase in spin-glass literature, where only a pure state exists. In this case, the dependence on the replica indices can be expressed in terms of three different sectors, the *longitudinal* (a scalar field), the *anomalous* (one-replica-index) and the *replicon* (tensorial field) sectors.

One might wonder why our interest is mostly focused on the replicon mode. It is essentially the most sensitive to the transition and is responsible for possible RSB solutions. A zero (replicon) mode of the stability matrix corresponds to have a pole in the propagator (using field-theory jargon) and then a diverging susceptibility [5]. The longitudinal mode instead can give information in terms of spinodal points, describing how a state opens up along an unstable direction and originates a saddle.

Thanks to the definition in Eq. (6.43), we can immediately see that the most relevant terms are the ones below:

$$
\begin{aligned}
\frac{\partial S}{\partial Q_{ab}} &= i\hat{Q}_{ab} + -\frac{P}{2}p(1-p)\langle\hat{\Delta}_a\hat{\Delta}_b\rangle_c \\
\frac{\partial^2 S}{\partial Q_{ab}\partial Q_{cd}} &= \frac{P^2}{4}p^2(1-p)^2\langle\hat{\Delta}_a\hat{\Delta}_b,\hat{\Delta}_c\hat{\Delta}_d\rangle_c \\
\frac{\partial S}{\partial\hat{Q}_{ab}} &= iQ_{ab} - \frac{i}{N}\langle g_a g_b\rangle_c \\
\frac{\partial^2 S}{\partial\hat{Q}_{ab}\partial\hat{Q}_{cd}} &= \frac{1}{N^2}\langle g_a g_b, g_c g_d\rangle_c \\
\frac{\partial S}{\partial m_a} &= i\hat{m}_a + Pip\langle\hat{\Delta}_a\rangle \\
\frac{\partial^2 S}{\partial m_a\partial m_b} &= -P^2 p^2\langle\hat{\Delta}_a\hat{\Delta}_b\rangle_c \\
\frac{\partial S}{\partial\hat{m}_a} &= im_a - i\langle g_a\rangle \\
\frac{\partial^2 S}{\partial\hat{m}_a\partial\hat{m}_b} &= \langle g_a g_b\rangle_c \,,
\end{aligned}
\tag{6.44}
$$

where the label c stands for the connected part of the correlation function, i.e. $\langle g_a g_b, g_c g_d\rangle_c \equiv \langle g_a g_b g_c g_d\rangle - \langle g_a g_b\rangle\langle g_c g_d\rangle$. Note that terms like $\frac{\partial^2 S}{\partial Q_{ab}\partial\hat{Q}_{cd}}$ or $\frac{\partial^2 S}{\partial m_a\partial\hat{m}_b}$ would only provide a constant factor. Then, we would like to get rid of the additional quantities $\hat{Q}$ and $\hat{m}$ and to write an expression for the functional S in terms of Q and m only. We consider the expression of the first derivative of the action with respect to the overlap matrix:

$$
\frac{dS}{dQ_{ab}} = \frac{\partial S}{\partial Q_{ab}} + \sum_{ef}\frac{\partial S}{\partial\hat{Q}_{ef}}\frac{d\hat{Q}_{ef}}{dQ_{ab}} + \sum_e\frac{\partial S}{\partial m_e}\frac{dm_e}{dQ_{ab}} + \sum_e\frac{\partial S}{\partial\hat{m}_e}\frac{d\hat{m}_e}{dQ_{ab}} \,.
\tag{6.45}
$$

For the second derivatives as well we have to take care of all mixing terms. Thanks to the saddle-point condition with respect to m, we can take advantage of this additional relation:

$$\frac{d}{dQ_{ab}}\frac{\partial S}{\partial m_c} = \frac{\partial^2 S}{\partial m_c \partial Q_{ab}} + \frac{\partial^2 S}{\partial m_c^2}\frac{dm_c}{dQ_{ab}} = 0 , \tag{6.46}$$

from which we get:

$$\frac{dm_c}{dQ_{ab}} = -\left(\frac{\partial^2 S}{\partial m_c \partial Q_{ab}}\right)\left(\frac{\partial^2 S}{\partial m_c^2}\right)^{-1} . \tag{6.47}$$

We will not show in detail the complicated expression of the second derivative. For that, we refer the interested reader to Appendix B. Once the full expression of the second derivatives of the action with respect to the overlap is available, we have to take two contributions into account: one coming from the diagonalization of the $\hat{\Delta}$-dependent correlation function and another from the g-dependent term. In the thermodynamic limit, the latter should not contribute to the replicon eigenvalue, hence reducing the computation to the following term:

$$\mathcal{M}_{ab,cd} \equiv \frac{\partial^2 S}{\partial Q_{ab}\partial Q_{cd}} = \frac{P^2}{4}p^2(1-p)^2\langle \hat{\Delta}_a\hat{\Delta}_b, \hat{\Delta}_c\hat{\Delta}_d\rangle_c . \tag{6.48}$$

Thanks to the symmetry of the replica indices, the matrix can be diagonalized in the replica symmetric phase as follows:

$$\mathcal{M}_{ab,cd} = \mathcal{M}_{ab,ab}\left(\frac{\delta_{ac}\delta_{bd} + \delta_{ad}\delta_{bc}}{2}\right) + \mathcal{M}_{ab,ac}\left(\frac{\delta_{ac} + \delta_{bd} + \delta_{ad} + \delta_{bc}}{4}\right) + \mathcal{M}_{ab,cd} . \tag{6.49}$$

The projection on the replicon sector then reads:

$$\lambda_{\mathrm{repl}} \equiv \mathcal{M}_{ab,ab} - 2\mathcal{M}_{ab,bc} + \mathcal{M}_{ab,cd} . \tag{6.50}$$

$$\lambda_{\mathrm{repl}} = \frac{P^2 p^2(1-p)^2}{4}\left(\langle \hat{\Delta}_a^2\hat{\Delta}_b^2\rangle - \langle \hat{\Delta}_a\hat{\Delta}_b\rangle^2 - 2\langle \hat{\Delta}_a^2\hat{\Delta}_b\hat{\Delta}_c\rangle + 2\langle \hat{\Delta}_a\hat{\Delta}_b\rangle\langle \hat{\Delta}_a\hat{\Delta}_c\rangle + \right.$$

$$\left. + \langle \hat{\Delta}_a\hat{\Delta}_b\hat{\Delta}_c\hat{\Delta}_d\rangle - \langle \hat{\Delta}_a\hat{\Delta}_b\rangle\langle \hat{\Delta}_c\hat{\Delta}_d\rangle\right) =$$

$$= \frac{P^2 p^2(1-p)^2}{4}\left(\langle \hat{\Delta}_a^2\hat{\Delta}_b^2\rangle - 2\langle \hat{\Delta}_a^2\hat{\Delta}_b\hat{\Delta}_c\rangle + \langle \hat{\Delta}_a\hat{\Delta}_b\hat{\Delta}_c\hat{\Delta}_d\rangle\right) . \tag{6.51}$$

We consider first the last term, i.e. the four-point correlation function with all different replica indices, as the same conclusions safely apply to the other two correlators.

A simple way to have a Gaussian integral and to decouple replicas is to introduce an auxiliary Gaussian variable z. The correlator $\langle \hat{\Delta}_a\hat{\Delta}_b\hat{\Delta}_c\hat{\Delta}_d\rangle$ would then correspond:

$$\frac{\int \prod_a \frac{d\Delta_a d\hat{\Delta}_a}{2\pi}\theta(-\Delta_a)e^{\left[i\sum_a \hat{\Delta}_a(\Delta_a+pm^*)-\frac{p(1-p)(\bar{q}-q)}{2}\sum_a \hat{\Delta}_a^2-\frac{p(1-p)q+\epsilon^2}{2}(\sum_a \hat{\Delta}_a)^2\right]}\hat{\Delta}_a\hat{\Delta}_b\hat{\Delta}_c\hat{\Delta}_d}{\int \prod_a \frac{d\Delta_a d\hat{\Delta}_a}{2\pi}\theta(-\Delta_a)e^{\left[i\sum_a \hat{\Delta}_a(\Delta_a+pm^*)-\frac{p(1-p)(\bar{q}-q)}{2}\sum_a \hat{\Delta}_a^2-\frac{p(1-p)q+\epsilon^2}{2}(\sum_a \hat{\Delta}_a)^2\right]}}$$

(6.52)

which, in terms of z, becomes:

$$\langle \hat{\Delta}_a\hat{\Delta}_b\hat{\Delta}_c\hat{\Delta}_d\rangle = \int \mathcal{D}z \frac{\left(\int \frac{d\Delta d\hat{\Delta}}{2\pi}\theta(-\Delta)e^{i\hat{\Delta}(\Delta+pm^*)-\frac{p(1-p)(\bar{q}-q)}{2}\hat{\Delta}^2+iz\sqrt{p(1-p)q+\epsilon^2}\hat{\Delta}}\hat{\Delta}\right)^4}{\left(\int \frac{d\Delta d\hat{\Delta}}{2\pi}\theta(-\Delta)e^{i\hat{\Delta}(\Delta+pm^*)-\frac{p(1-p)(\bar{q}-q)}{2}\hat{\Delta}^2+iz\sqrt{p(1-p)q+\epsilon^2}\hat{\Delta}}\right)^4}.$$

(6.53)

The equation above represents the average over z of the conditioned probability of the four-replica-index correlator, conditioned to z. By integrating out the conjugated field $\hat{\Delta}$, the denominator D, turns out to be:

$$D = \int \frac{d\Delta\,\theta(-\Delta)}{\sqrt{2\pi}}\frac{e^{-\frac{(\Delta+pm^*+z\sqrt{p(1-p)q+\epsilon^2})^2}{2p(1-p)(\bar{q}-q)}}}{\sqrt{p(1-p)(\bar{q}-q)}} = \frac{1}{2}\left\{1 + \text{Erf}\left(\frac{m^*p+z\sqrt{p(1-p)q+\epsilon^2}}{\sqrt{2}\sqrt{p(1-p)(\bar{q}-q)}}\right)\right\}.$$

(6.54)

A similar expression can be written for the other two terms in Eq. (6.51). Instead of embarking on long and hard computations, we can immediately realize that the three pieces appearing in the expression for the replicon mode are powers or appropriate combinations of Eq. (6.54). Thanks to this simplification, we can rewrite the three correlators, still functions of the Gaussian variable z, as:

$$\langle \hat{\Delta}_a^2\hat{\Delta}_b^2\rangle_z - 2\langle \hat{\Delta}_a^2\hat{\Delta}_b\hat{\Delta}_c\rangle_z + \langle \hat{\Delta}_a\hat{\Delta}_b\hat{\Delta}_c\hat{\Delta}_d\rangle_z = \left(\frac{\partial^2 D}{\partial z^2}\frac{1}{\psi^2 i^2 D^2}\right)^2 +$$
$$- 2\left(\frac{\partial^2 D}{\partial z^2}\frac{1}{\psi^2 i^2 D^2}\right)\left(\frac{\partial D}{\partial z}\frac{1}{\psi i D}\right)^2 + \left(\frac{\partial D}{\partial z}\frac{1}{\psi i D}\right)^4,$$

(6.55)

where we used i to indicate the imaginary unit and the additional notations[8]:

$$\psi = \sqrt{p(1-p)q+\epsilon^2}, \quad b = p(1-p)(\bar{q}-q), \quad c = pm^*,$$

(6.56)

As mentioned in the introduction of this model, the parameter ψ represents the variance of the resource surplus.

Let us see in detail what the shielded and the vulnerable phases imply in terms of the leading eigenvalue of the stability matrix, as plotted in Fig. 6.4 first. Its expression is shown below:

---

[8]The label * means that a saddle-point computation has been performed over m.

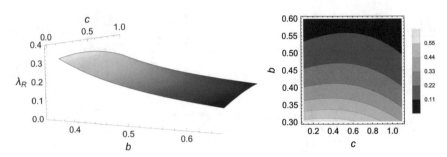

Fig. 6.4 On the left, three-dimensional plot of the replicon eigenvalue at finite ψ, as a function of the two parameters b and c as defined in the main text. On the right, a contour plot at fixed ψ. Figures taken from [1]

$$
\lambda_{\text{repl}} \propto \int \mathcal{D}z \frac{4e^{-\frac{2(c+\psi z)^2}{b}}\left[b + 2\sqrt{2\pi b}(c+\psi z)e^{\frac{(c+\psi z)^2}{2b}} + 2\pi(c+\psi z)^2 e^{\frac{(c+\psi z)^2}{b}}\right]}{b^3\pi^2\left[1+\text{Erf}\left(\frac{c+\psi z}{\sqrt{2b}}\right)\right]^4}.
$$

(6.57)

The replicon mode has in principle a complicated expression, which can nevertheless be simplified or numerically investigated in certain limits. Indeed, upon approaching the transition at $\epsilon = 0$, the parameter $\psi = \sqrt{p(1-p)q + \epsilon^2}$, which actually drives the transition, goes to zero linearly. It can be shown that in this regime a vanishing replicon mode appears.[9] Summarizing, our computation should be regarded as an analytical confirmation of the emergence of a marginally stable phase, which exactly coincides with the *shielded* phase, where $\psi = 0$ (Fig. 6.5).

Note that we are looking only at positive values of c, even though in principle there is no constraint on this parameter. For negative c possible singularities can emerge and a more detailed analysis would be needed. Another interesting extension to work on in the future concerns a possible breaking phase, which would be signaled by a negative replicon mode. In principle, since the Lyapunov function is convex and allows us to map this setting onto the perceptron in the convex regime, a replica symmetry breaking cannot occur. In particular, this scenario appears to be more related to a *random linear programming* problem [8], which consists in slightly modifying the original perceptron model by relaxing the spherical constraint. In that case, the RS solution continues to hold everywhere, but it is marginally stable for some specific values of the control parameter α.

---

[9]We nevertheless remind that the sharp S/V transition takes place only for $\epsilon \to 0$, otherwise it would be replaced by a crossover regime.

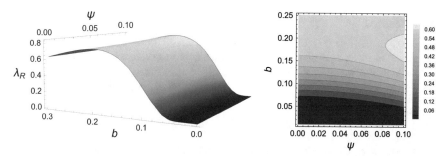

Fig. 6.5 On the left, a three-dimensional representation of the solution of the numerical integration in Eq. (6.57) as a function of ψ and b at fixed and positive c. The transition to the shielded phase is determined by a vanishing parameter $\psi = \sqrt{p(1-p)q + \epsilon^2}$ as $\epsilon \to 0$. In the limit $a, b \to 0$ the replicon eigenvalue touches zero. On the right, we show an isoline plot in the same parameter space at fixed c

6.6 Conclusions

This chapter presents work at the interface between two different fields that, in principle, can appear uncorrelated: amorphous systems close to the jamming transition and ecology. Our goal is to reduce the *gradient term* as much as possible. Keeping in mind all the discussions on the perceptron model, we studied a high-dimensional ecosystem model, on which the mathematical formalism and the physical interpretations seem to converge to very close conclusions. Some preliminary numerical tests have been done in parallel to reproduce the probability distributions of the availabilities and the species abundances in the different phases associated with the model, which are not presented here to give more emphasis on the analytical framework. In particular, we managed to derive the asymptotic behavior of the spectral density, which appeared as a useful tool to distinguish between a shielded and a vulnerable phase. The relative deviations in the two phases can be better interpreted in terms of zero modes. We managed to prove the emergence of a marginally stable phase, having derived the expression of the stability matrix and its leading eigenvalue. The replicon mode, which drops to zero while approaching the S phase, is a signature of an underlying isostaticity condition. A different scenario occurs in the V phase, where the replicon mode is generally finite.

We presented just a first-order computation, as the starting point for further and more complex analysis. Despite the fact we considered a simple replica symmetric (RS) approximation, we would be interested in evaluating the presence of a possible RSB phase and then the emergence of multiple attractors in the phase space. This might have various and important consequences for the dynamics, as mentioned in the introduction for the Lotka-Volterra-like equations and extensively analyzed in [2, 4].

References

1. Altieri A, Franz S (2019) Constraint satisfaction mechanisms for marginal stability and criticality in large ecosystems. Phys Rev E 99(1):010401(R)
2. Biroli G, Bunin G, Cammarota C (2018) Marginally stable equilibria in critical ecosystems. New J Phys 20(8):083051
3. Braun M (1975) Differential equations and their applications. Springer
4. Bunin G (2017) Ecological communities with Lotka-Volterra dynamics. Phys Rev E 95(4):042414
5. De Dominicis C, Giardina I (2006) Random fields and spin glasses. Cambridge University Press
6. Fischer KH, Hertz JA (1991) Spin glasses. Cambridge Studies in Magnetism
7. Fisher CK, Mehta P (2014) The transition between the niche and neutral regimes in ecology. Proc Natl Acad Sci 111(36):13111
8. Franz S, Parisi G (2016) The simplest model of jamming. J Phys A: Math Theor 49(14):145001
9. Kessler DA, Shnerb NM (2015) Generalized model of island biodiversity. Phys Rev E 91(4):042705
10. Lotka AJ (1926) Elements of physical biology. Sci Progress Twent Century (1919–1933) 21(82):341
11. Lyapunov AM (1992) The general problem of the stability of motion. Int J Control 55(3):531
12. Mac Arthur R (1969) Species packing, and what competition minimizes. Proc Natl Acad Sci 64(4):1369
13. Marchenko VA, Pastur LA (1967) Distribution of eigenvalues for some sets of random matrices. Matematicheskii Sbornik 114(4):507
14. Neill WE (1974) The community matrix and interdependence of the competition coefficients. Am Nat 108(962):399
15. Odum EP (1959) Fundamentals of ecology. WB Saunders Company
16. Tikhonov M (2016) Community-level cohesion without cooperation. Elife 5:e15747
17. Tikhonov M, Monasson R (2017) Collective phase in resource competition in a highly diverse ecosystem. Phys Rev Lett 118(4):048103
18. Tilman D (1982) Resource competition and community structure. Princeton University Press
19. Volterra V (1927) Variazioni e fluttuazioni del numero d'individui in specie animali conviventi. C. Ferrari
20. Wigner EP (1993) Characteristic vectors of bordered matrices with infinite dimensionsi. In: The Collected Works of Eugene Paul Wigner. Springer

Part II
Lattice Theories Beyond Mean-Field

Chapter 7
The M-Layer Construction

This chapter attempts to develop a systematic perturbative formalism around the Bethe solution. The results presented here are the object of a recent paper [3] with collaborators from the Universities of Rome and Turin. This work was motivated by long discussions started in the first year of my Ph.D. After several months we decided to come back to this topic to properly investigate the role of finite-size corrections in fully-connected as well as in Bethe lattice models.

Despite the large applicability of mean-field approximations in statistical physics, a big limitation is due to their inadequacy to describe real systems. Much effort has been devoted in the last decades to achieve a deeper understanding of diluted models. Albeit more demanding than fully-connected models, diluted models mimic in an intuitive and straightforward way finite-dimensional systems, as they are both characterized by a finite connectivity. Furthermore, in this class of models local observables associated with one (or more) spins depend on their neighborhood and have nontrivial distributions, while in the fully-connected version only global observables are important.

In this context, the Bethe approximation [21] is an essential tool that provides analytic results for a wide class of problems. Given its conceptual and practical equivalence with the Belief Propagation algorithm [34], it is also quite important in modern computer science, such as for error-correction codes and optimization problems [9]. Clearly, it is just an approximation and to have a good description, especially in terms of critical phenomena, it is important to consider the presence of loops. Several studies have thus been proposed to include the effect of loops in a lattice associated with the given problem, starting from an expansion around a mean-field-like solution [9, 22, 28, 31, 31, 34]. Such methods look very promising although they were conceived for single realizations of the disorder and not to average over different samples: in a sense, they cannot be used to distinguish between annealed and quenched averages. Other approaches attempted to improve the consistency of this approximation using linear response theorem [30], which, again, fails in the presence of loops. One of the main difficulty is that in many interesting problems the Bethe approximation is non-perturbative, in the sense that there is no small parameter that can be used to develop an expansion to compute higher-order corrections [11, 32].

© Springer Nature Switzerland AG 2019
A. Altieri, *Jamming and Glass Transitions*, Springer Theses,
https://doi.org/10.1007/978-3-030-23600-7_7

To address all these issues we will propose in following the M-layer construction, valid for both ordered and disordered systems. It consists in replicating M times the original model.[1] Then, for each link of the original model, we will have M copies of it connecting sites on the same layers. We will then rewire those links creating inter-layer connections with some random permutations. We will show that in the large M-limit the Bethe approximation, which generally fails on non-tree-like topologies, becomes asymptotically exact. The quantity $1/M$ will play the role of the small parameter here, used to build a perturbative expansion around the Bethe approximation. This kind of construction was actually introduced some years ago by Vontobel [33] in the computer science literature. Nowadays the computation of a rigorous $1/M$ expansion [23, 24] is still an active research field. Differently than the preceding works in computer science, we are motivated by a different purpose. We aim to develop a reliable formalism as a tool to study critical phenomena in finite-dimensional systems.In particular, we will consider the large M limit, focusing on a restricted region near the Bethe critical point, the *critical region*, where deviations from mean-field behavior are expected.[2]

The need to go beyond mean field stems from the fact that there are important second-order phase transitions that display essential differences (or they are not present at all) between the Bethe lattice (namely an infinite tree where each vertex has fixed connectivity) and the corresponding fully-connected (FC) version. This is the case for spin glasses in an external field. In the FC model there is a transition line in the field-temperature plane, the so-called de Almeida-Thouless line [26], which ends at infinite field at zero temperature. This implies that there is no transition in field at $T = 0$ and the system is always in the SG phase where replica symmetry is broken. The same argument does not hold in finite dimensions, where it is known that at $T = 0$ and high enough field no spin-glass phase occurs. No clear proof of the existence of a spin-glass transition at finite field exists although it is the subject of ongoing research [4, 5]. Since the Bethe lattice displays a transition at finite field h_c in the zero temperature regime [27], one might think to perform an expansion around this model to understand the fate of the transition in finite dimensions.

A similar situation occurs for disordered models of glasses. They are described by a one-step Replica Symmetry Breaking (1RSB) theory, displaying a discontinuous phase transition in the FC model. However, many of them, rephrased in finite dimensions, can either display a continuous transition more similar to the spin-glass behavior in a field, or do not display a transition at all. This happens already in the Bethe lattice, since the finite connectivity of the lattice can change or entirely destroy the nature of the transition [8].

---

[1] In the case of a system with quenched random disorder, as random magnetic fields, the disorder is chosen with the same distribution for all the M copies.

[2] In the modern theory of second-order phase transitions, critical exponents are evaluated starting from continuum field theories rather than microscopic models [15, 19, 25, 35]. This is typically justified invoking universality. However, a more direct connection between microscopic models and continuum field theories can be established by a closely related construction: the fully-connected M-layer [7, 15, 35].

Another example is given by the Anderson localization. The existence of this type of transition is prohibited by definition in the FC case: if each site is linked to all the others, a localized state cannot exist. Things are different in the Bethe lattice where, thanks to its finite connectivity, a localization transition does exist [1, 6]. However, in finite dimension neither the exact solution for such transition nor its upper critical dimension are known.

In some other cases, the transition in the FC model is present but with significant differences with respect to its finite-dimensional version. This is the case of the Random Field Ising model (RFIM). In this case, the FC mean-field solution and the corresponding loop expansion is well understood [10]. This expansion implies dimensional reduction, meaning that the critical exponents of RFIM in d dimensions are the same of a pure ferromagnet in $d' = d - 2$ dimensions [29]. However, it is well-known that for low enough dimensions, dimensional reduction is no longer valid because of non-perturbative effects. For the RFIM, one can write a self-consistent equation for the local magnetization, which has a unique solution only in the FC case. The assumption of a unique solution is the necessary condition to preserve the validity of dimensional reduction. Conversely, in the case of finite connectivity, multiple solutions exist, a property shared by both finite-dimensional lattices and Bethe lattices. One possibility is that the loop expansion around the FC fixed point leads to wrong results precisely because it does not account correctly for the presence of multiple solutions: therefore, expanding around the Bethe solution, which displays a fixed point with several solutions, could give different results [29].

The lack of a proper method on a FC model implies the lack of a well-established continuum field theory to be eventually studied by a loop expansion. As we will see in the following, the Bethe *M*-layer construction allows us to give a contribution in this field and to obtain critical exponents, even without knowing the underlying continuum field theory of the given problem.

7.1 The Curie-Weiss Model

In this section, we will discuss how the Landau-Ginzburg Hamiltonian can be derived from the Ising model. In a sense, this represents an intermediate step to address the problem of a loop expansion in field theory.

In a very general way, we first consider the paradigmatic ferromagnetic transition in the fully-connected Curie-Weiss model. Starting from the standard Ising model on the lattice (where $\sigma_i = \pm 1$) and performing a marginal modification on the interaction term of the nearest-neighbor Hamiltonian, the problem can be solved exactly. The Curie-Weiss model, where each variable interacts with order N other spins, is defined by:

$$\mathcal{H}_{CW}(\sigma) = -\frac{J}{2N} \sum_{i,j}^{1,N} \sigma_i \sigma_j - h \sum_{i=1}^{N} \sigma_i . \tag{7.1}$$

In zero field the ground-state energy per spin is equal to $-J/2$ compared to the value $-Jd$, which corresponds to the nearest-neighbor Ising ferromagnet on a d-dimensional lattice. Making use of the Gaussian identity to rewrite the canonical partition function, we get:

$$Z = \sum_{\sigma_i} \frac{1}{\pi} \int_{-\infty}^{+\infty} dx \, e^{\left[-x^2+2\left(\frac{\beta J}{2N}\right)^{1/2} x \sum_{i=1}^{N} \sigma_i + \beta H \sum_{i=1}^{N} \sigma_i \right]} =$$

$$= \frac{1}{\pi} \int_{-\infty}^{+\infty} dx \, e^{-x^2} \left\{ 2 \cosh \left[2 \left(\frac{\beta J}{2N}\right)^{1/2} x + \beta H \right] \right\}^N .$$

(7.2)

The expression above can be simply rewritten changing variables $2 \left(\frac{\beta J}{2N}\right)^{1/2} x = \beta J m$, namely:

$$Z \propto \left(\frac{\beta J N}{2\pi}\right)^{1/2} \int_{-\infty}^{+\infty} dm \, e^{-N\beta f_\beta(m,h)} ,$$

(7.3)

where the action reads:

$$f_\beta(m,h) = \frac{1}{2} J m^2 - \frac{1}{\beta} \log \left[2 \cosh(\beta J m + \beta h) \right] .$$

(7.4)

As the free energy is now an extensive quantity, we can perform a saddle-point computation on it, which leads to the Curie-Weiss equation of state:

$$m = \tanh(\beta J m + \beta h) .$$

(7.5)

In the case of a zero external field, this self-consistent equation for the magnetization admits only the solution $m = 0$ for $\beta \leq \beta_c = 1/J$, while for $\beta > \beta_c$ two additional solutions with opposite sign emerge, as a signature of the ferromagnetic phase.

To study finite-size corrections to the FC solution one can perform a systematic expansion starting from Eq. (7.4). In principle this procedure will lead to different results with respect to the standard Ginzburg-Landau theory, although we will see that the resulting loop expansion will reduce to the standard theory around the critical fixed point. Universality will be preserved in any case.

For simplicity we will focus on $h = 0$. Since at the critical point $\beta_c = 1/J$, the magnetization of the system is zero, thus allowing us to expand Eqs. (7.3) and (7.4) around the critical value $m = 0$ and obtain:

$$Z \propto \int_{-\infty}^{+\infty} dm \cdot e^{-N\left(\tau m^2 + g_4 m^4 + g_6 m^6 + \dots\right)}$$

(7.6)

where we defined the reduced temperature $\tau = \frac{\beta J}{2}(1 - \beta J)$ and the coupling constants $g_4 = \frac{\beta^4 J^4}{12}$, $g_6 = -\frac{\beta^6 J^6}{45}$. The Lagrangian of this problem can be interpreted as composed by a Gaussian free part $\mathcal{L}_0 = N\tau m^2$, plus an interaction term: $\mathcal{L}_{int} = \sum_{i=2}^{\infty} g_{2i} m^{2i}$, having vertices of all even orders. The free propagator $G_0^{(2)}$, that is the expectation value of the field m^2 on the free part, is just $G_0^{(2)} = \frac{1}{N\tau}$. To better understand the effect of the interaction part on the propagators we start writing simple Feynman rules. In the expansion of G^E, that is the correlator between E external points, we can have all possible graphs constructed in the following way:

- Insert vertices of order $k = 4, 6, 8, \ldots$, each k-vertex emanating k lines;
- Connect each line either to other lines or to external points, excluding connections that generate vacuum diagrams;
- Associate to each line the free propagator $G_0^{(2)}$;
- Associate to each k-vertex a factor $N g_k$.

Thus, taking a generic graph with E external lines, I internal lines, V vertices and L loops, its behavior as a function of the dimension N will be: $N^{-I-E+V} = N^{-L-E+1}$, where we used the known relation[3]: $L = I - V + 1$ [19]. This means that, for a fixed number of external lines, the $1/N$ expansion reduces to an expansion in the number of loops. Let us see in detail what it means for the two-point correlators, namely for $E = 2$. To have the propagator at the first order in the $\frac{1}{N}$-expansion, we should have $N^{-L-E+1} = N^{-1}$, which implies $L = 0$. This order corresponds to the free propagator $G_0^{(2)} = \frac{1}{N\tau}$ without loops. At order $\frac{1}{N^2}$ we should have all the contributions with $L = 1$: in this case we have just one diagram, with a $k = 4$-vertex, that gives a contribution $G_1^{(2)} = \frac{N g_4}{(N\tau)^3} = \frac{1}{N^2} \frac{g_4}{\tau^3}$. At order $\frac{1}{N^3}$ two loops are needed. We can then construct two diagrams, with two vertices of order $k = 4$, giving a contribution $G_2^{(2)a,b} = \frac{(N g_4)^2}{(N\tau)^5} = \frac{1}{N^3} \frac{g_4^2}{\tau^5}$ each, and a diagram with a $k = 6$-vertex, giving a contribution $G_2^{(2)c} = \frac{N g_6}{(N\tau)^4} = \frac{1}{N^3} \frac{g_6}{\tau^4}$.

The relevant graphs at a given order in $1/N$ can have different k-vertices. This is deeply different from what happens, for instance, in the standard ϕ_4 Ginzburg-Landau theory for a ferromagnetic model, where just $k = 4$ vertices are present. The situation can, however, change in proximity of the critical point. Let us look more in detail what that means.

At a given order in the $\frac{1}{N}$-expansion, we will have different contributing diagrams whose temperature evolution will be: $\tau^{-E-I} = \tau^{-L-V-E+1} = \tau^{-L-\frac{2I}{k}-E(1+\frac{1}{k})+1}$. For a k-vertex, we made use of the following relation: $k \cdot V = 2I + E$. This means that at the same order in $\frac{1}{N}$, i.e. at fixed L, and at fixed E, diagrams with generic k-vertices are all divergent when $\tau \to \infty$. However, the most divergent diagrams are those with

---

[3]A method to derive the above relation $L = I - V + 1$ is the following: (i) by cutting an internal line one also remove one loop, thus $L - I$ is constant; (ii) a diagram without loops has a tree-like structure, meaning that if one removes one vertex on the boundary of the tree implicitly one converts an internal line into an external line and again $I - V$ becomes constant. In the Fourier representation, when $D > 0$, the number of loops corresponds also to the number of free momenta on which one integrates. According to the momentum conservation constraint in each vertex, such a number must be equal to the number of propagators minus the number of vertices plus one to guarantee the total momentum conservation.

smaller k, that is with $k = 4$: therefore, near the critical point we recover the standard ϕ_4 theory, as expected according to the universality principle.

Once the relevant class of diagram is known, that is for $k = 4$, let us zoom into them. Their contribution is $\frac{N^{V-I-E}}{\tau^{I+E}} = \frac{N^{-\frac{1}{2}-\frac{3}{4}E}}{\tau^{I+E}} \propto \frac{1}{(N\tau^2)^{I/2}}$. For a fixed number of external lines, we can identify a critical region, where $N\tau^2 = O(1)$ and the finite-size corrections do contribute. Looking close enough to the transition, i.e. for $\tau < \frac{1}{\sqrt{N}}$, the mean-field description is no longer valid. This corresponds to a kind of generalized Ginzburg criterion.

We can now introduce new scaling variables: $\tilde{m} = N^{1/4}m$, $\tilde{\tau} = \sqrt{N}\tau$ that, once inserted in Eq. (7.6), allow us to eliminate the dependence on N in the first terms, i.e.:

$$Z \propto \int_{-\infty}^{+\infty} dm \cdot e^{-\tilde{\tau}\tilde{m}^2 - g_4 \tilde{m}^4 - g_6 \frac{1}{\sqrt{N}} \tilde{m}^6 + \dots}. \tag{7.7}$$

This rescaling shows explicitly that higher k-vertices are sub-dominant. The finite size N destroys the transition present in the thermodynamic limit and the divergences in the physical observables at the critical point are replaced by peaks at finite N, whose expression is independent of N in the rescaled variables. The advantage is that in the critical region we do not need to evaluate the whole integral, but only to retain the quadratic and the quartic term for the leading corrections. Interestingly, one might consider a different model, e.g. a soft-spin version, where clearly the coupling constants and the series would be different outside the critical region. However, provided that all variables are correctly rescaled, in the critical region all models are well-described by the same critical theory.

7.2 M-Layer Expansion Around Fully-Connected Models

To establish a reasonable perturbative expansion, which can give predictions in finite dimensions, we consider a model defined on a d-dimensional lattice and essentially replicate the system M times. This construction is also known as M-layer model. Let us consider the paradigmatic Ising model on a lattice, defined by the following Hamiltonian:

$$H = -\frac{1}{2} \sum_{i,j=1}^{N} J_{ij} \sigma_i \sigma_j - \sum_{i=1}^{N} h_i \sigma_i , \tag{7.8}$$

where the matrix J defines the structure of the lattice, being one or zero depending on whether i and j are nearest neighbours. On each site of the lattice we insert a stack of M Ising spins. If i and j are nearest neighbours on the original lattice, in the M-layer construction each spin σ_i^α will interact with all the $\sigma_j^{\alpha'}$, where α and α' run from 1 to M. This means that now each spin is coupled with M other spins on each of its $2D$ nearest neighbors. Each variable is thus labelled by a couple of indices (i, α), where i runs over the number of lattice sites $i = 1, ..., N$ and α runs over the layers $\alpha = 1, ..., M$.

In principle the sum in Eq. (7.8) contains M^2 terms and since the Hamiltonian must be proportional to M we need to rescale everything by $1/M$. This rescaling guarantees the correct $M \to \infty$ limit. Then the partition function turns out to be:

$$Z_M = \sum_{\{\sigma_i^\alpha\}} \exp\left\{ \frac{\beta}{2M} \sum_{i,j=1}^{N} J_{ij} \sum_{\alpha,\alpha'=1}^{M} \sigma_i^\alpha \sigma_j^{\alpha'} + \beta \sum_{i=1}^{N} h_i \sum_{\alpha=1}^{M} \sigma_i^\alpha \right\}. \tag{7.9}$$

Then, we make use of the *Hubbard-Stratonovich transformation* to rewrite the partition function in a more straightforward way. The sum over the spin variables can be done immediately and yields:

$$Z \propto \int_{-\infty}^{+\infty} \prod_{i=1}^{N} dm_i \, \exp\left\{ M\left[-\frac{\beta}{2} \sum_{ij} m_i (J^{-1})_{ij} m_j + \sum_i \log\left[2\cosh(\beta m_i + \beta h_i) \right] \right] \right\} \tag{7.10}$$

where each layer gives the same contribution to the partition function. We are interested in the $M \to \infty$ limit, that is the mean-field version of the fully-connected M-layer construction. In this case, the integral is dominated by the value that maximizes the expression at the exponent, in agreement with the result of Sect. 7.1. On the other hand, for $M = 1$ we recover the original model. We can thus reinterpret the $1/N$ expansion around the fully connected model, explained in Sect. 7.1, inside the M-layer construction: the precedent expansion is exactly the $1/M$-expansion in the $d = 0$ case, where just a point is considered in the original lattice. However, introducing a spatial structure has non-trivial implications now. Again, once a suitable rescaling of all quantities as functions of $1/M$ is done, the correspondence between the loop expansion and the expansion in $1/M$ can be clearly understood. In a diagrammatic approach, the propagator is $O(1/M)$ while the vertices have a contribution $O(M)$. Taking into account the relation linking the number of internal lines I, the number of vertices V and loops L, i.e. $L = I - V + 1$, one can easily compute the order of each diagram in terms of $M^{V-I} \sim (1/M)^{L-1}$. Therefore, a Feynman diagram with L loops is proportional to $1/M^L$. As already mentioned, the expansion in $1/M$, for large M,[4] is formally equivalent to the expansion in $1/N$ above: the difference now is that the divergent contributions are also due to the integration in momentum space, which has to be treated in a more subtle way than in the Curie-Weiss model.

---

[4]Indeed, the mean-field approximation for $M = 1$ is not very accurate.

7.2.1 The Propagator in Momentum Space Near the Criticality

The ordinary Feynman rules continue to hold, allowing us to associate diagrams to a perturbative formalism. These rules are a direct consequence of Wick's theorem [19], namely:

- we assign to each vertex a factor $-g$;
- to each line joining two points x_i and x_j we assign a propagator $G_0(x_i - x_j)$;
- we integrate over all internal points z_i, i.e. $\int d^d z_i$;
- to each diagram we associate a multiplicative prefactor, the symmetry factor [14], which counts all possible permutations of lines at fixed vertex preserving the structure of the diagram.

The same rules also apply in momentum space, with the only difference that to each independent loop corresponds now an integration like $\int \frac{1}{(2\pi)^d} d^d q$. The integrations associated with independent loops generate a product of δ functions as the condition of the overall momentum conservation has to be fulfilled in each of them.

Then, to obtain the expression of the propagator in momentum space, let us introduce an infinitesimal external field $h \sim O(\epsilon)$ allowing non-trivial solutions, and expand the equation of state (7.5) near the critical value $m = 0$. The saddle-point equation for the magnetization becomes:

$$m_i = \tanh\left(\beta \sum_k J_{ik} m_k + \beta H_i\right) \approx \beta\left(\sum_k J_{ik} m_k + H_i\right), \qquad (7.11)$$

or equivalently:

$$H_i = \frac{m_i}{\beta} - \sum_k J_{ik} m_k = \sum_k \left(\frac{\delta_{ik}}{\beta} - J_{ik}\right) m_k = \sum_k B_{ik} m_k \qquad (7.12)$$

from which $m_i = \sum_k A_{ik} H_k$ and $A_{ik} \equiv \left(B^{-1}\right)_{ik}$. Its Fourier transformation implies:

$$\hat{B}(\vec{k}) = \frac{1}{\beta}\left(1 - 2\beta \sum_{\mu=1}^{d} \cos(\vec{k}_\mu)\right) \qquad (7.13)$$

where, by a matrix inversion, the propagator is given by $\hat{G}_0(k) = \frac{1}{\beta}\hat{B}^{-1}(k) = \left(1 - 2\beta \sum_{\mu=1}^{d} \cos(k_\mu)\right)^{-1}$. As we are interested in small-momentum behavior, we can make use of the following expansion: $\cos(k) = 1 - \frac{k^2}{2} + \frac{1}{4!}k^4 - \frac{1}{6!}k^6 + O(k^8)$, which in d dimensions implies:

$$\hat{G}_0(k) = \frac{1}{1 - \frac{1}{d}\sum_{\mu=1}^{d}\cos(k_\mu)} = \left(1 - 2d\beta + \beta\sum_{\mu=1}^{d}k_\mu^2 + O(k_\mu^4)\right)^{-1}. \quad (7.14)$$

Then the critical temperature is simply given by $\beta_c = 1/(2d)$. In principle, one should retain all terms in the expansion of $\cos(k_\mu)$, but again near the critical point one can neglect all the other terms except for the first one, once a proper rescaling of all quantities of interest is done. Indicating as $\tau = (1 - 2d\beta)$ and redefining $k_\mu^2 = (\tau/\beta)\tilde{k}_\mu^2$, the leading term in Eq. (7.14) is: $\left(1 + \sum_\mu k_\mu{}^2 - \frac{\tau}{12\beta}\sum_\mu k_\mu{}^4 + \frac{1}{360}\frac{\tau^2}{\beta^2}\sum_\mu k_\mu{}^6 + ...\right)$. Close to the critical temperature k_μ^4 and higher-order terms do not contribute. We can then conclude that in the critical region near the mean-field critical temperature the leading divergences are the same that would be obtained in a loop expansion starting from a Ginzburg-Landau theory. Clearly, a direct connection between the behavior in momentum space and in direct space can be established, leading to similar conclusions for k and m if one treats the problem perturbatively.

7.2.2 A Continuum Field Theory

To better understand the leading and subleading contributions in the critical region, we can evaluate the action in momentum space as a function of $\epsilon = 4 - d$, namely for an ordinary quartic theory. Similar studies, based on a perturbative renormalization group approach, have been performed in a multicomponent cubic theory near the critical temperature [2] to evaluate the leading behavior in $6 - \epsilon$ dimensions of *composite operators* [19, 35].

When dealing with a continuum field theory, it might be convenient to introduce the coarse-grained variable $m(x) = \frac{1}{V}\sum_{i\in x}\sigma_i$. According to that, the Hamiltonian of a ferromagnetic system can be rewritten as: $-J\sum_{\langle ij\rangle}\sigma_i\sigma_j = \frac{J}{2}\sum_{\langle ij\rangle}(\sigma_i - \sigma_j)^2 -$ const. $\sim \frac{a^2J}{2}\sum_{\langle ij\rangle}\frac{(\sigma_i-\sigma_j)^2}{a^2}$, where a is the lattice spacing. In the continuum limit a gradient term appears, i.e. $\vec{\nabla}m(x)$, which penalizes widely fluctuating configurations. The generating functional in direct space in the presence of a quartic vertex then reads:

$$Z_M = \int \mathcal{D}m(x)\exp\left[-\beta M\int d^d x\left(\frac{\hat{a}^2}{2}(\vec{\nabla}m(x))^2 + \frac{\tau(\beta)}{2}m^2(x) + \frac{g}{4!}m^4(x)\right)\right]. \quad (7.15)$$

Adsorbing the dependence on β and integrating by parts, we can write more compactly:

$$Z_M = \int \mathcal{D}m(x)\exp\left[-\frac{M}{2}\int d^d x\, m(x)\left(-\vec{\nabla}^2 + \tau\right)m(x) - \frac{gM}{4!}\int d^d x\, m^4(x)\right]. \quad (7.16)$$

As explained in the previous section for the $1/N$ expansion, a proper rescaling in order to eliminate the dimensional dependence on the Gaussian term, i.e. $\tilde{m}(x) = M^{1/2}m(x)$, is useful to recognize as most divergent contributions the diagrams associated with the lowest order vertex, i.e. $\frac{1}{M}\tilde{m}^4(x)$. Then, we only need to compute the contributions coming from one-loop diagrams, characteristic of a m^4-theory, while to the order $1/M^2$ we should also consider the contributions of two-loop diagrams, typical of a m^6-expansion. This is why we can only take one coupling constant g, differently than the generic case, where the expressions are functions of g_m, each of them associated with a different power of m. Introducing the following rescalings:

$$x = b_x \tilde{x} \qquad \tau = b_\tau \tilde{\tau} \qquad m = b_m \tilde{m} \,, \tag{7.17}$$

we can write a set of equations to be satisfied simultaneously:

$$\begin{cases} b_\tau b_m^2 b_x^d = 1/M \\ b_m^2 b_x^{d-2} = 1/M \\ b_m^4 b_x^d = 1/M \,. \end{cases} \tag{7.18}$$

The first and the third ones are just the generalization of analogous relations written for $d = 0$. The second relation is instead necessary when $d > 0$ and the propagator contains a momentum term. Recalling the generic N-point correlation function $G^{(N)}(k)$ in momentum space:

$$(2\pi)^d \delta^{(d)} \left(\sum_{i=1}^N k_i \right) G^{(N)}(k_i) = \int \prod_{i=1}^N d^d x_i\, e^{ik_i x_i}\, G^{(N)}(x_i) \,, \tag{7.19}$$

we get the following set of equations where $b_m \to b_m b_k^{-d}$:

$$\begin{cases} b_\tau b_m^2 b_k^{-d} = 1/M \\ b_m^2 b_k^{-2d} b_k^{d+2} = 1/M \\ b_m^4 b_k^{-4d} b_k^{3d} = 1/M \,. \end{cases} \tag{7.20}$$

Note that, as expected, $b_k = b_x^{-1}$. Moreover, the integration in momentum space is extended over the interval $[0, \Lambda]$, which properly comes from the observable definition on a lattice.[5] After a few algebraic passages we eventually obtain these critical exponents:

$$\begin{cases} \alpha_\tau = \frac{2}{d-4} \\ \alpha_m = \frac{1}{d-4} \\ \alpha_x = -\frac{1}{d-4} \end{cases} \tag{7.21}$$

---

[5]Clearly, integrating up to the cut-off Λ is equivalent to stating that no further information on scales smaller than the linear size of the cells is available.

where we defined $b_* \equiv M^{\alpha_*}$. The exponents above are in agreement with the scalar case mean-field predictions, i.e for $d = 0$, and also with the usual argument in critical phenomena indicating $d = 4$ as the upper critical dimension of the model. The exponents change sign in $d = 4$ meaning that above this critical dimension the theory is no longer renormalizable and an analytical continuation is not feasible. Note that, according to the rescaling $m = M^{\alpha_m} \tilde{m}$, the coefficient α_m should be negative to guarantee a well-defined expansion, where small fluctuations are under control. In the same philosophy we have to rescale the cut-off as $\Lambda = b_k \tilde{\Lambda}$ and then to rewrite the partition function as:

$$Z_M \propto \int_{-\infty}^{+\infty} \prod_{|k|<\Lambda} d\tilde{m}(k) d\tilde{m}^*(k) \exp\{M S(\tilde{m}(k))\}. \tag{7.22}$$

The action $S(\tilde{m}(k))$ takes the form:

$$S(\tilde{m}(k)) = -\int d^d k \ (\tau + c_2 k^2 + c_4 k^4 + \ldots)|\tilde{m}(k)|^2 - g_4 \sum_i m_i^4 + \ldots \tag{7.23}$$

where $\tilde{m}(k)$ is the Fourier transform of $\{m_i\}$ and Λ is the momentum cut-off, i.e. the inverse of the lattice spacing. For $d < 4$ we can introduce the following rescalings:

$$x M^{-\frac{1}{\epsilon}} \to x, \ \tau M^{\frac{2}{\epsilon}} \to \tau, \ m M^{\frac{1}{\epsilon}} \to m, \tag{7.24}$$

where $\epsilon \equiv 4 - d$. We obtain an expression that does not depend on M, plus small corrections:

$$Z_M \propto \int_{-\infty}^{+\infty} \prod_k d\tilde{m}(k) d\tilde{m}^*(k) \ \exp\left[-\int d^d k \ (\tau + c_2 k^2)|\tilde{m}(k)|^2 - g_4 \int d^d x \ m(x)^4 + \right.$$

$$\left. - \sum_{s>4} g_s \ O\left(M^{-\frac{s-4}{\epsilon}}\right) - \sum_{s>2} c_s \ O\left(M^{-\frac{s-2}{\epsilon}}\right) \right] \tag{7.25}$$

where we have used $k M^{\frac{1}{\epsilon}} \to k$ and $\tilde{m} M^{-\frac{d-1}{\epsilon}} \to \tilde{m}$. The rescaled cut-off was sent to infinity, obtaining in this way the continuum limit of the Ginzburg-Landau theory. Summarizing, the fully connected M-layer construction has the following properties:

- In the large M limit at *fixed* temperature the mean-field approximation is accurate and the loop expansion is perturbative. In this region, all coupling constants are relevant for the $1/M$ expansion.
- Deviations from a mean-field behavior are observed in a region centered at the mean-field critical temperature that shrinks with M as $1/M^{2/\epsilon}$. Correlation functions are Gaussian at short distances also in the critical region, while a non-trivial behavior is observed at large distances. More precisely any n-point correlation function at large distances, $O(M^{1/\epsilon})$, and with small deviations from the mean-field critical temperature, $O(1/M^{2/\epsilon})$, behaves as $1/M^{n/\epsilon}$ with a scaling function

that is determined by the Ginzburg-Landau theory in the continuum limit. Therefore, only the quartic coupling constant turns out to be relevant.

The study of the Ginzburg-Landau Hamiltonian in the continuum limit leads to
the well-known problem of renormalization. As the continuum limit is not well
defined for $d \geq 4$, the above scalings are meaningless for $d > 4$, while in $d = 4$
they just tell us that the size of the region, where deviations from the mean-field-like
behavior appear, vanishes (since $\epsilon = 0$). The continuum limit is instead well-defined
in $d < 4$. However, if we try to compute correlations by a loop expansion we discover
that all coefficients of the series are divergent at large values of τ, due to the cut-off
divergence. This effect is intrinsically related to the continuum limit. Therefore, the
renormalization procedure provides the reshuffling tools to obtain reasonable results
[19, 25, 35].

7.3   The Bethe Approximation

We aim at generalizing the arguments previously applied to fully-connected models to lattice theories where the *Bethe approximation* becomes a fundamental tool.
We provide hints in order to better define this approximation and derive the final
expression for the Bethe free energy.

There exist different ways to express graphically the structure of interactions of
random variables, one of these consisting in the factor graph formalism [16, 17, 21].
One of the main problems is computing marginal distributions of one variable or
joint distributions of a set of random variables. The update rules giving exact results
for the marginals on tree-like graphs have been discovered independently in different
contexts: in statistical physics it is usually called the *Bethe Peierls approximation*,
in coding theory the *sum product* algorithm, while in computer science is the *belief
propagation*, always denoted as BP.

Let us define a graphical model with N random variables $\vec{\sigma} = (x_1, x_2, ..., x_N)$.
The joint probability distribution of the $\vec{\sigma}$'s variables has the form:

$$P(\vec{\sigma}) = \frac{1}{Z} \prod_{a=1}^{M} \psi_a(\vec{\sigma}_{\partial a}) , \qquad (7.26)$$

where $\vec{\sigma}_{\partial_a} \equiv \{\sigma_i | i \in \partial a\}$, given that $\partial a \subseteq [N]$. ψ_a are called *compatibility functions*,
non-negative, and taking values on real space. Each edge (i, a) is defined by a variable
node i and a function node a. The advantage of dealing with tree-like graphs is that
after t iterations the messages asymptotically converge to their fixed-point values:
$\nu_{i \to a}^{\infty}(\sigma_i)$, which corresponds to the single-variable marginal, where the node a has
been pruned and $\hat{\nu}_{a \to i}^{\infty}(\sigma_i)$ which is the analogous complementary message assuming
that all factors ∂i, adjacent to i except a, have been removed.

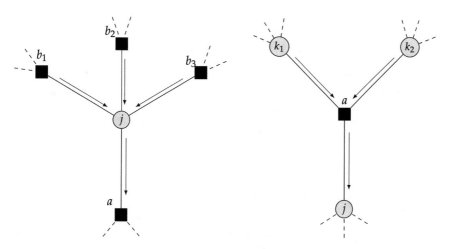

Fig. 7.1 On the left a part of a factor graph referred to the message $\hat{\nu}_{b \to j}(\sigma_j)$. On the right the corresponding part of the factor graph involved in the computation of $\hat{\nu}_{a \to j}(\sigma_j)$, function of the messages $\nu_{k \to a}(\sigma_k)$

The belief propagation method requires the computation of the following rules (as shown in Fig. 7.1):

$$
\begin{cases}
\nu_{j \to a}(\sigma_j) \simeq \displaystyle\prod_{b \in \partial j \setminus a} \hat{\nu}_{b \to j}(\sigma_j) \\[2mm]
\hat{\nu}_{a \to j}(\sigma_j) \simeq \displaystyle\sum_{\vec{\sigma} \in \partial a \setminus j} \psi_a(\vec{\sigma}_{\partial a}) \prod_{k \in \partial a \setminus j} \nu_{k \to a}(\sigma_k)
\end{cases}
\tag{7.27}
$$

where the symbol "$\simeq$" means equality up to a normalization factor. Using all incoming messages one can immediately estimate the single-variable marginal distribution of i, namely:

$$
P(\sigma_i) \simeq \prod_{a \in \partial i} \hat{\nu}_{a \to i}(\sigma_i) .
\tag{7.28}
$$

Therefore, the internal energy is directly computable as the expectation value of the energy of the whole system, i.e.:

$$
\begin{aligned}
U &= -\sum_{\vec{\sigma}} P(\vec{\sigma}) \sum_{a=1}^{M} \log \psi_a(\vec{\sigma}_{\partial a}) = \\
&= -\sum_{a=1}^{M} \frac{1}{Z_a} \sum_{\vec{\sigma}_{\partial a}} \left(\psi_a(\vec{\sigma}_{\partial a}) \log \psi_a(\vec{\sigma}_{\partial a}) \prod_{i \in \partial a} \nu^*_{i \to a}(\sigma_j) \right) ,
\end{aligned}
\tag{7.29}
$$

and the normalization factor Z_a is:

$$Z_a = \sum_{\vec{\sigma}_{\partial a}} \psi_A(\vec{\sigma}_{\partial a}) \prod_{i \in \partial a} \nu^*_{i \to a}(\sigma_j) \qquad (7.30)$$

∗ denoting the fixed-point value of the message. Note that, as already mentioned, these equations provide the correct result only for a tree-like graph. In the presence of loops the BP equations do not converge.

Once the distribution $P(\vec{\sigma})$ is known, the entropy can be written as:

$$S[P] = -\sum_{\sigma} P(\vec{\sigma}) \log P(\vec{\sigma}) , \qquad (7.31)$$

and for a tree-graphical model $P(\vec{\sigma})$ becomes:

$$P(\vec{\sigma}) = \prod_{a \in \mathcal{F}} P_a(\vec{\sigma}_{\partial a}) \prod_{i \in \mathcal{V}} P_i(\sigma_i)^{1-|\partial i|} \qquad (7.32)$$

$\mathcal{F}$ and $\mathcal{V}$ being the subsets of edges and vertices, respectively. Since the entropy reads:

$$S[P] = -\sum_{a \in \mathcal{F}} P_a(\vec{\sigma}_{\partial a}) \log P_a(\vec{\sigma}_{\partial a}) - \sum_{i \in \mathcal{V}} (1 - |\partial i|) P_i(\sigma_i) \log P_i(\sigma_i) \qquad (7.33)$$

and by definition the free energy is $\mathcal{F}[P] = U[P] - TS[P]$, we have all the tools to write the Bethe-Peierls free energy as a function of local marginals:

$$\mathcal{F}\left(\{\nu\}, \{\hat{\nu}\}\right) = \sum_{a \in \mathcal{F}} \mathcal{F}_a\left(\{\nu\}\right) + \sum_{i \in \mathcal{V}} \mathcal{F}_i\left(\{\hat{\nu}\}\right) - \sum_{(i,a) \in E} \mathcal{F}_{ia}\left(\{\nu\}, \{\hat{\nu}\}\right) . \qquad (7.34)$$

In the thermodynamic limit the free energy of a diluted system exactly corresponds to the Bethe free-energy. Conversely, if the number of nodes N is finite, the average free-energy density accounts for emerging loopy structures as well. Recent results have been made possible by performing a $1/N$ expansion of the free energy, in both a Random Regular Graph (RRG) [20] and an Erdös-Rényi Graph, an inhomogeneous locally tree-like graph (ERG) [12]. Within this approach first $O(1/N)$ corrections come from the average number of loops of definite length multiplied by the free energy shift, which is due to the addition of non-intersecting paths to an infinite tree-like graph. However, its weakness is due to the impossibility to generalize and extend this method close to the critical point. The next terms in the $1/N$ expansion diverge and the terms that were sub-leading in the replica-symmetric phase become relevant in this critical regime.

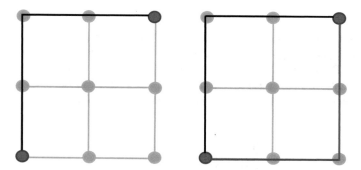

Fig. 7.2 In the 0-th order of our Bethe expansion, we consider the sum of all linear paths as if they were on a Bethe lattice, namely as if the neighboring spins were independent. An example of such a path is shown above on the left. At the 1-th order of the expansion, there are loops. The simplest ones are those formed by two single lines crossing, such as those shown on right

7.3.1 A Simple Argument Behind the Bethe Expansion

This Section is devoted to explaining the intuition behind the Bethe-like expansion. We consider, for instance, an Ising model in which we aim to compute the connected correlation function between the two red dots in Fig. 7.2 on a two-dimensional lattice:

$$\langle \sigma_i \sigma_j \rangle_c = \frac{\sum_{\{\sigma\}} \sigma_i \sigma_j e^{-\beta H}}{\sum_{\{\sigma\}} e^{-\beta H}} - \frac{\sum_{\{\sigma\}} \sigma_i e^{-\beta H}}{\sum_{\{\sigma\}} e^{-\beta H}} \cdot \frac{\sum_{\{\sigma\}} \sigma_j e^{-\beta H}}{\sum_{\{\sigma\}} e^{-\beta H}} . \tag{7.35}$$

The main difficulty comes from the fact that we have to simultaneously sum over all the lattice spins, such spins being correlated to all the others. The advantage of mean-field calculations, both fully connected or Bethe ones, is that the sums over neighboring spins can be factorized. What invalidates this argument in finite dimension is the presence of loops. With this in mind, we can think to approximate the correlation function we are interested in by partitioning the path in single mean-field-like pieces. In the first step of this approximation, we assume spins i and j to be connected by single lines, in which all the spins are independent of the others. This is true if the link between them on that particular line is removed, as is the case on a Bethe lattice. One of these paths is represented in the left part of Fig. 7.2. At this point, the correlation between i and j can be approximated by the sum of the correlations on the independent lines, called l, which are easy to compute once the solution on the Bethe lattice model is known. Then the 0-th order term in this approximation will be:

$$\langle \sigma_i \sigma_j \rangle^0 = \sum_{l \in \{\text{lines}\}} \langle \sigma_i \sigma_j \rangle_l . \tag{7.36}$$

However, one can object that spins on the lines connecting i and j are not independent because of the presence of loops in the lattice. In the right part of Fig. 7.2

we can see two lines—considered as independent in the 0-th order approximation of our expansion—which generate a single loop. To refine our previous estimate, we can imagine taking spatial loops into account for the computation of correlation functions. The simplest loops are those identified by two lines, say l_1 and l_2, that meet in two points. We explained how to compute observables on Bethe lattices where we have simple loops like these. Thus the first-order correction for the correlation function will be:

$$\langle \sigma_i \sigma_j \rangle^1 = \sum_{(l_1,l_2)\in\{\text{single loops}\}} \langle \sigma_i \sigma_j \rangle_{l_1,l_2} - \langle \sigma_i \sigma_j \rangle_{l_1} - \langle \sigma_i \sigma_j \rangle_{l_2} , \qquad (7.37)$$

where the subtractions eliminate the 0-th order contribution of lines l_1 and l_2 considered as independent. This approximation can be further refined by considering loops formed by three, four and more lines.

7.4   Mathematical Formalism for a Hamiltonian System on a Bethe Lattice

The aim of this Section is to recall standard techniques that will be used on the Bethe lattice. Generalizations of these results can be safely applied in the thermodynamic limit to RRGs and ERGs.

The Hamiltonian of a general model with one and two-body interactions can be written as:

$$\mathcal{H}(\{\sigma_i\}) = -\sum_{(i,j)} J(\sigma_i, \sigma_j) - \sum_i H(\sigma_i) . \qquad (7.38)$$

The present treatment is valid for any type of variables $\{\sigma_i\}$, not only for Ising spins. We assume a factorized prior on the variables, i.e. $\prod_i d\mu(\sigma_i)$.[6] Systems with quenched disorder can be also discussed in this framework by means of the replica method. In this case, σ_i are replicated variables, but after averaging over disorder the Hamiltonian becomes homogenous.

Under some ergodicity assumption on the Gibbs measure, namely the replica symmetry, one can recursively find the solution $Q(\sigma)$—that we assume to be unique—of the following implicit equation:

$$Q(\sigma) = \frac{1}{Z_Q} \int d\mu(\tau) \, e^{\beta J(\sigma,\tau)+\beta H(\tau)} \, Q(\tau)^{c-1} , \qquad (7.39)$$

where c is the connectivity degree on the Bethe lattice and

---

[6]For instance, the Ising case corresponds to $d\mu(\sigma_i) = \delta(\sigma_i + 1) + \delta(\sigma_i - 1)$.

$$\mathcal{Z}_Q = \int d\mu(\sigma)\, d\mu(\tau)\, e^{\beta J(\sigma,\tau)+\beta H(\tau)}\, Q(\tau)^{c-1} \tag{7.40}$$

is the partition function. The distribution $Q(\sigma)$ is called the *cavity distribution*, satisfying the condition $\int d\mu(\sigma)\, Q(\sigma) = 1$ by construction. Gathering all the information above, the thermodynamics of the model can be solved. Then, the marginal probability distribution of a variable in the infinite Bethe lattice is given by:

$$P(\sigma) \propto Q(\sigma)^c\, e^{\beta H(\sigma)} . \tag{7.41}$$

To compute two-point correlators among variables at finite distance L on the Bethe lattice, we notice that they are exactly the same as the correlations on a one-dimensional system of length L with the same couplings $J(\sigma,\tau)$ and an effective field on the internal variables, which is due to $c-2$ cavity fields in addition to H. Therefore, to solve the one-dimensional model, we introduce the (unnormalized) cavity fields

$$\rho_k(\sigma) \equiv Q(\sigma)^k\, e^{\beta H(\sigma)} \tag{7.42}$$

and the symmetric transfer matrix

$$T(\sigma, \tau) \equiv \sqrt{\rho_{c-2}(\sigma)}\, e^{\beta J(\sigma,\tau)}\, \sqrt{\rho_{c-2}(\tau)} . \tag{7.43}$$

The transfer matrix can be written in terms of its orthonormal eigenvectors as:

$$T(\sigma, \tau) = \sum_\lambda \lambda\, e_\lambda(\sigma) e_\lambda(\tau), \tag{7.44}$$

with $\int d\mu(\sigma)\, e_\lambda(\sigma) e_\gamma(\sigma) = \delta_{\lambda,\gamma}$. Using Eq. (7.39) one can see that

$$\sqrt{\rho_{c-2}(\sigma)}\, Q(\sigma) \tag{7.45}$$

is actually an eigenvector of the transfer matrix. According to the Perron-Frobenius theorem we can argue that this eigenvector corresponds to the largest eigenvalue, $\lambda_{max} = \mathcal{Z}_Q$, and then we have:

$$e_{\lambda_{max}} \propto \sqrt{\rho_c(\sigma)} . \tag{7.46}$$

The marginal probability is given exactly by its squared value:

$$P(\sigma) = e^2_{\lambda_{max}}(\sigma) , \tag{7.47}$$

where the normalization of the probability follows from the normalization of the eigenvector. Since we are interested mainly in two-point correlations we consider a closed one-dimensional chain of length N and we then take the thermodynamic limit $N \to \infty$. From this we obtain:

$$P_L(\sigma, \tau) = \frac{\sum_{\lambda, \gamma} \gamma^{N-L} \lambda^L e_\gamma(\sigma) e_\lambda(\sigma) e_\gamma(\tau) e_\lambda(\tau)}{\sum_\lambda \lambda^N} \tag{7.48}$$

and we notice that in the thermodynamic limit both terms proportional to N are dominated by the largest eigenvalue, namely:

$$P_L(\sigma, \tau) = \sum_\lambda \left(\frac{\lambda}{\lambda_{\max}}\right)^L e_{\lambda_{\max}}(\sigma) e_\lambda(\sigma) e_{\lambda_{\max}}(\tau) e_\lambda(\tau) . \tag{7.49}$$

Introducing the following rescaled eigenvalues:

$$\frac{\lambda}{\lambda_{\max}} \to \lambda \tag{7.50}$$

and the functions:

$$a_\lambda(\sigma) \equiv e_{\lambda_{max}}(\sigma) e_\lambda(\sigma) , \tag{7.51}$$

we can thus write:

$$P_L(\sigma, \tau) = P(\sigma) P(\tau) + \sum_{|\lambda|<1} \lambda^L a_\lambda(\sigma) a_\lambda(\tau) , \tag{7.52}$$

where $\int d\mu(\sigma) a_\lambda(\sigma) = 0$ for all $\lambda < 1$, invoking again the Perron-Frobenius theorem. We will use this result in the following Sections to reproduce the computation on the M-layer model.

From the exact expression of the two-point distribution one can also obtain an expression for two-point susceptibilities on the Bethe lattice, i.e.:

$$\chi(A) \equiv \sum_{i \neq 1} \left(\langle A(\sigma_1) A(\sigma_i) \rangle - \langle A(\sigma_1) \rangle \langle A(\sigma_i) \rangle \right) \tag{7.53}$$

as

$$\chi(A) = c \sum_{L=1}^{\infty} (c-1)^{L-1} \sum_{|\lambda|<1} \lambda^L \left(\int d\mu(\sigma) \, a_\lambda(\sigma) A(\sigma) \right)^2$$
$$= c \sum_{|\lambda|<1} \frac{\lambda}{1 - \lambda/\lambda_c} \left(\int d\mu(\sigma) \, a_\lambda(\sigma) A(\sigma) \right)^2 \tag{7.54}$$

where the critical eigenvalues is $\lambda_c \equiv \frac{1}{c-1}$. As the largest eigenvalue—that is usually smaller than 1—approaches the critical value λ_c, the susceptibility diverges, provided that $A(\sigma)$ has a non-zero projection on the corresponding function $a_\lambda(\sigma)$.

7.5 The Bethe M-Layer

After a brief explanation of the formalism to be employed on the Bethe lattice, we can now consider the Bethe M-layer construction. The model we describe here is inspired by the M-layer construction around the fully-connected model, already shown in Sect. 7.2. Again, we consider an original model on a finite-dimensional lattice, and we replicate it M times. For the sake of simplicity, we focus on the ferromagnetic Ising model but the results we are going to present can be safely extended to a variety of other models. Each variable on this multi-layer model is labelled by a couple of indices (i, α), where i runs over the number of lattice sites ($i = 1, ..., N$) and α over the layers ($\alpha = 1, ..., M$). In the original lattice the only existing couplings are those between spins on the same layer: $J_{i,j}^{\alpha,\beta} \neq 0$ iff $\alpha = \beta$. This is represented in the left part of Fig. 7.3. At this point, for each couple (i, j) of neighbouring spins belonging to the original graph, one can define a permutation π of the links $J_{i,j}^{\alpha,\pi_\beta}$ that automatically creates inter-layer connections. This intermediate step of permuting links randomly is called *rewiring*. An example of a possible permutation of the links between two neighboring spins is shown in the central part of Fig. 7.3. Hence, the Hamiltonian of the M-layer construction can be written as:

$$H = - \sum_{(i,j)=1}^{N} \sum_{\alpha,\beta=1}^{M} \sigma_i^\alpha \sigma_j^\beta J_{ij}^{\alpha,\beta} C_{ij}^{\alpha,\beta} , \tag{7.55}$$

where $C_{ij}^{\alpha,\beta}$ is the adjacency matrix, whose elements are induced by the uniformly chosen permutation of the links. Their joint probability distribution is factorized on each edge (i, j):

$$\mathcal{P}(C) = \prod_{(i,j)} P(C_{ij}),$$

with

$$P(C) = \frac{1}{M!} \sum_{\{C^{\alpha,\beta}=0,1\}} \prod_\alpha \delta \left(\sum_\beta C^{\alpha\beta} = 1 \right) \prod_\beta \delta \left(\sum_\alpha C^{\alpha\beta} = 1 \right).$$

The main difference with respect to the M-layer around the FC model relies on the fact that now the connectivity of the model remains that of the original one, namely $z = 2D$ on a hypercubic d-dimensional lattice. The free energy of the model is defined as the quenched average over possible permutations of the links. If the original model is disordered, to each layer we associate a different realization of the disorder, where the couplings $J_{i,j}^\alpha$ are i.i.d. random variables. Then the free energy of the model is defined as the quenched average over both the disorder and the permutations.

Considering a finite M essentially corresponds to introducing loopy structures in the model. The generation of a certain kind of loops is shown in the rightmost

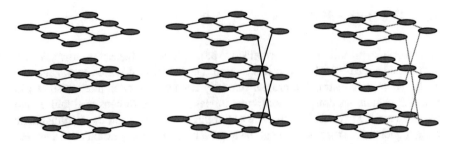

Fig. 7.3 Qualitative representation of a M-layer construction for a $2D$ regular lattice with $M = 3$. (Left) the original lattice replicated M times; (center) a possible rewiring of its M copies; (right) the emergence of a single loop corresponding to a non-backtracking closed path when projected onto the original lattice

part of Fig. 7.3. In a critical phenomena perspective we will study the divergence of the $1/M$ corrections at each order as we approach the Bethe critical temperature, with the goal of using these series to extract the non-mean-field critical exponents in finite dimension. Analogously to what shown in Sect. 7.2, the free energy in the limit $M \to \infty$ can be found through a saddle-point computation and it coincides with the free-energy of the Bethe model with the same connectivity as the original lattice.

The loop expansion defined in this formalism involved loops that represent now the spatial ones, while the bare correlators are those computed in the Bethe approximation. If one aims to study disordered systems, for which non-perturbative effects are crucial, this new expansion might really give important insights into the physics of finite-dimensional models.

7.6 $1/M$ Corrections by the Cavity Method

We consider two distinguished vertices on the lattice, x and y and the connected correlation function between two spins σ_x^α and σ_y^β where the Greek indices label the M layers. This correlation is a random quantity depending on the realization of the M-lattice. The key point of our analysis is that in the large M limit the given realization of the M-lattice is locally tree-like with probability one. Thus, for any x and y, the two spins are not correlated in the $M \to \infty$ limit.

Small corrections to the average value at finite M are due to the rare re-wirings in which the two spins turn out to be close to each other. The simplest possibility is when spins are connected by a path of finite length L. At leading order the probability that a given path of length L is present is given by M^{-L}, as every link of the path is present with probability $1/M$. On the other hand, there are many such paths. To correctly count them, it is more convenient to order them according to their projection on the original lattice. Given a path on the original lattice between x and y, the total number of paths connecting σ_x^α and σ_y^β is M^{L-1} at leading order since each of the $L - 1$ internal vertices can be chosen in M different layers. Therefore, the total weight of the paths having the same projection on the original lattice is $1/M$. More precisely one should sum over non-backtracking paths [13] on the original lattice thanks to the fact that, given one site (x, α) of the layered graph, there is only one value of $\beta = 1, \ldots, M$ such that the site $(x + \mu, \beta)$ is connected to (x, α). A backtracking path is defined as follows: in the projection on the original lattice, vertex repetitions are allowed but conditioned to the fact that three consecutive vertices are pairwise different.

Therefore, defining $P(\sigma_x^\alpha, \sigma_y^\beta)$ as the marginal distribution over spins σ_x^α and σ_y^β for a given M graph realization, which is obtained by tracing the Boltzmann distribution over all other spins, we can write the connected distribution as:

$$P_c(\sigma_x^\alpha, \sigma_y^\beta) \equiv P(\sigma_x^\alpha, \sigma_y^\beta) - P(\sigma_x^\alpha)P(\sigma_y^\beta) , \qquad (7.56)$$

while the average over re-wirings is given by:

$$\langle P_c(\sigma_x^\alpha, \sigma_y^\beta)\rangle_{rew} = \frac{1}{M} \sum_{L=1}^{\infty} \#_L(x - y) P_{c,L}(\sigma_x^\alpha, \sigma_y^\beta) + O\left(\frac{1}{M^2}\right) . \qquad (7.57)$$

$P_{c,L}(\sigma, \tau)$ represents the connected correlation between two sites at distance L on a Bethe lattice, while the symbol # stands for the total number of non-backtracking paths of length L between point x and point y.[7] We refer the interested reader to Appendix C for more details concerning the computation of non-backtracking paths.

According to the formalism introduced in Sect. 7.4, we can decompose $P_{c,L}(\sigma, \tau)$ and re-express it in terms of the functions $a_\lambda(\sigma)$, which are related to the eigenvectors of a transfer matrix. We thus have:

$$P_{c,L}(\sigma, \tau) = \sum_{\lambda < 1} \lambda^L a_\lambda(\sigma) a_\lambda(\tau) \qquad (7.58)$$

provided that $\int d_\mu(\sigma)\sigma a_\lambda(\sigma) = 0$ for $\lambda \neq 1$. If we consider the disconnected probability we must add a term $P(\sigma)P(\tau)$ that corresponds to the eigenvalue 1. This allows us to write the above expression in terms of the generating function of the non-backtraking path:

$$B_\lambda(x - y) = \sum_{L=1}^{\infty} \#_L(x - y)\lambda^L \qquad (7.59)$$

and then:

$$\langle P_c(\sigma_x^\alpha, \sigma_y^\beta)\rangle_{rew} = \frac{1}{M} \sum_{\lambda \neq 1} B_\lambda(x - y) a_\lambda(\sigma_x^\alpha) a_\lambda(\sigma_y^\beta) + O\left(\frac{1}{M^2}\right) . \qquad (7.60)$$

---

[7] On the Bethe lattice correlations are associated with one-dimensional chains and can be analyzed by transfer matrix methods.

For the Ising model without disorder the situation is much simpler, having only one eigenvalue besides $\lambda = 1$. In disordered models one can nevertheless take advantage of the replica formalism.

7.6.1 Critical Behavior

Upon approaching a second-order phase transition on the Bethe lattice the second largest eigenvalue λ' approaches $\lambda_c \equiv 1/(2D-1)$. To better investigate the critical behavior, we can consider the Fourier transform of Eq. (7.60) and concentrate on the limit of small k's close to the critical point, where only the second largest eigenvalue is relevant. We can then neglect all other eigenvalues and write:

$$\langle P_c(\sigma, \tau, k) \rangle_{rew} \approx \frac{1}{M} \sum_{\lambda \neq 1} \hat{B}_\lambda(k) a_{\lambda'}(\sigma) a_{\lambda'}(\tau) , \qquad (7.61)$$

where $\hat{B}_\lambda(k)$ stands for the Fourier transform of $B_\lambda(x - y)$. On the other hand, the behavior for small k of the Fourier transform of the generating function of non-backtracking paths is precisely given by $B_\lambda(k) = (\lambda - \lambda_c + k^2 + O(k^4))^{-1}$. Near the critical point the expression above reduces to:

$$\langle P_c(\sigma, \tau, k) \rangle_{rew} \approx \frac{1}{M} \frac{1}{m^2 + k^2} a_\lambda(\sigma) a_\lambda(\tau) , \qquad (7.62)$$

where $m^2 \propto \lambda' - \lambda_c$ is a linear function of the external parameters vanishing at the critical point.[8] We can stress once more that near the critical point the correlation in the M-layer construction has the same expansion at leading order as that for a fully-connected model: the reason essentially lies in the universality.

7.6.2 The Graph-Theoretical Expansion

Once an expression for the correlator has been obtained, we can attempt to write a reasonable and general diagrammatic expansion. Let us consider a generic physical process defined on a lattice with two-body interactions. We will mostly work without specifying the actual nature of the process but in practice we will assume that at least it is possible to study it on a Bethe lattice. This includes, of course, statistical mechanics models (with or without quenched disorder) at finite temperature but also systems where there is no Hamiltonian, e.g. percolation, k-core percolation or zero-temperature systems. Looking at a problem that is solvable on the Bethe lattice implies that the generic observable of order k can be computed on the lattice where

---

[8]We are implicitly employing here a field theory notation.

the k-points are the leaves of a tree-like structure embedded into an infinite tree. Similarly, we assume that the problem is also solvable (with a little bit more effort) if the lattice topology is such that a few loops are present but nevertheless the lattice extends to infinity intrinsically characterized by a tree-like structure.

Our goal is to compute observables as determined by these process. Again we do not need to specify in many details the nature of these observables but we expect them to satisfy the clustering property meaning that the two-point observable $O(x, y)$ factorizes onto the product of two one-point observables $O(x, y) \approx O'(x)O''(y)$ if the positions x and y are far away on the lattice. For this reason, we shall consider connected observables vanishing if one of its arguments becomes sufficiently large. We also assume that the generic observable can be written as a sum of products of connected observables of less or equal order. For a statistical mechanics problem with variables σ_x on the nodes of the lattice an example of a two-point observable is the probability $\langle\sigma_x\sigma_y\rangle$ induced by the Gibbs measure and the corresponding connected observable is $\langle\sigma_x\sigma_y\rangle - \langle\sigma_x\rangle\langle\sigma_y\rangle$. For a system with quenched disorder examples of two-point observables are $\overline{\langle\sigma_x\sigma_y\rangle}$ and $\overline{\langle\sigma_x\rangle\langle\sigma_y\rangle}$ and the corresponding connected observables are obtained subtracting the product $\overline{\langle\sigma_x\rangle\langle\sigma_y\rangle}$. We use here the standard notation, indicating as $\langle\bullet\rangle$ the thermal average and as $\overline{\bullet}$ the average over the quenched disorder.

Hence, if we have to evaluate an observable of order k on the lattice, we can study directly and independently tree-like pieces that extend to infinity. This results graphically in pruning all tree-like parts of the graph and in eventually dealing with a *finite* graph. At this point, we can introduce the notion of *fat diagram*. According to our definition, a fat diagram of order k satisfies the following properties:

- there are k distinguished vertices of any degree (including zero and one): we call them *external vertices*;
- additional vertices must have a degree equal or higher than two: these correspond to the *internal vertices*;
- each internal vertex must be connected at least to one of the external ones.

Note that a priori one cannot know what the internal and external vertices are. We first draw diagrams with both vertices and lines of any possible degree, under the assumption that internal and external vertices are obviously distinct. Then, for each specific diagram we choose a vertex labeling, i.e. whether a vertex is either internal or external[9] according to the above-mentioned rules.

Then, this definition encompasses a set of graphs with k external vertices, no dangling edges and no fully-disconnected components. Examples of fat diagrams are shown in Fig. 7.4. The first three examples represent two-point connected correlation functions, while the last one in the upper line provides an example of disconnected

---

[9]Topologically speaking, a given diagram can take part in the computation of both two-point and three-point correlations functions. We mostly focus in this work on two-point correlation functions. However, the choice of internal and external vertices is defined by the author after drawing the specific graph and defining the observables of interest. We will explain at the end of this chapter the importance of defining a proper internal vertex factor. This is really useful to understand why, for instance, a given diagram can contribute to a ϕ^3 theory but not to a ϕ^4 theory.

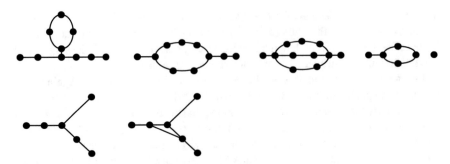

Fig. 7.4 Some examples of fat diagrams: in the first line, fat diagrams with two external lines, used to evaluate two-point correlation functions. In the second line we instead show two schematic examples with three external lines

functions. We shall see that they give a vanishing or a finite contribution depending on the observable we are looking at.

The idea of a topological expansion in terms of fat diagrams was proposed about thirty years ago by Efetov [11] and re-proposed in a more recent formulation by Parisi and Slanina [28], albeit with some technical difficulties. The derivation performed by Efetov in a supersymmetric formulation consisted in writing an effective action whose Feynman diagrams originated from the partition function of the original model plus that of the same model but further wrapped around it. Clearly, in the thermodynamic limit the wrapping does not play any relevant role. Interestingly, the saddle-point equations of the effective field theory defined in this way correspond to the Belief-Propagation equations, without any limitation in the use of this method. It can be applied to very general systems as long as defined on a lattice.

We now aim to deepen the properties of such fat diagrams. For a given fat diagram we define the observable $O(G)$ as the value that the observable takes on a graph that has the structure of the fat diagram and extends to infinity in a tree-like fashion, mimicking the local structure of the original lattice. Given the observable $O(G)$ we now introduce the notion of *line-connected* observable implicitly through the condition that the value of the observable on a fat diagram G is equal to the sum of the line-connected observables on all fat diagrams G' contained in G, namely:

$$O(G) = \sum_{G' \in G} O_{lc}(G') . \tag{7.63}$$

Line-connected observables can be determined iteratively starting from the simplest fat diagrams (tree-like) and then considering an increasing number of loops.

Now it is clear that the exact value of the observable of the problem is given by $O(G)$, G coinciding with the whole lattice. On the other hand, Eq. (7.63) applied to the whole lattice means that we can write $O(G)$ as a sum over all fat diagrams of the original lattice. Let us clarify better this point. The use of fat diagrams seems rather arbitrary given that we do not know if the original lattice resembles or not to a Bethe lattice. Indeed, if we sum the whole finite series the contributions of all fat diagrams other than the full lattice sum up to zero. On the other hand, we may ask what is the error obtained by summing only a subset of fat diagrams. The problem is that the contributions of different fat diagrams seems to be of the same order of magnitude and neglecting some of them affects the result significantly. More precisely we expect that the error is small only if we consider all possible fat diagrams of size up to the correlation length. Thus all truncation schemes are expected to fail close to a critical point where the correlation length diverges. In other words, the graph theoretical expansion is exact but non-perturbative: in this sense it is similar to the loop expansion in field theory. However, we can obtain a perturbative series considering the M-layer construction, as the fully-connected M-layer construction makes the field-theoretical loop expansion a perturbative expansion. In the case of the M-layer we can show that, although the exact result is only obtained summing all the terms of the series, if we consider a partial sum of all fat diagrams with a number of loops up to L, the error decreases with $1/M^L$.

7.6.3 The Graph-Theoretical Expansion on the M-Lattice

Our purpose is thus to evaluate a certain observable of order k on the M-lattice, averaged over all possible re-wirings. On each re-wiring of the M-lattice we can apply the exact formula (7.63). The average can then be written as:

$$\langle O \rangle_{rew} = \sum_G P(G) O_{lc}(G) . \tag{7.64}$$

The above expression means that we are summing over all possible fat diagrams G of the M-lattice, each weighted with the probability over the rewirings that the given fat diagram occurs.

It is convenient to classify all fat diagrams on the M-lattice according to their vertical projection on the original lattice. This is because the value of the observable $O(G)$ (and thus $O_{lc}(G)$) is the same for all fat diagrams with the same vertical projection. We recall that here we are not making any assumption of periodicity or homogeneity of the original lattice, actually proposing a totally general result. We can thus write:

$$\langle O \rangle_{rew} = \sum_G {}' W(G) O_{lc}(G) , \tag{7.65}$$

where the prime means that the sum is in the space of projections on the original lattice. The weight $W(G)$ is the sum over all fat diagrams of the M-lattice with vertical projection equal to G, each weighted with its probability $P(G)$. Note that the space of projections is larger than the space of fat diagrams on the original lattice: fat diagrams on the original lattice are such that each site and bond can be occupied only once, differently than projected fat diagrams for which we can have multiple occupancies of sites and bonds. The only constraint is that the lines of the projected fat diagrams must be non-backtracking paths on the original lattice.

Now we restrict the discussion of the weight $W(G)$ to connected diagrams G (see the first three examples in Fig. 7.4), under the assumption that O is a connected observable and it vanishes on a disconnected fat diagram. We consider first the case in which the projection is such that no site of the original lattice is occupied more than once. In this case, one can see that the sum over all possible realizations on the M lattice times the corresponding probabilities is exactly $1/M^{L+k-1}$, where L is the number of loops of G and k is the number of external vertices of the graph. Conversely, if the projection is such that some sites on the original lattice are occupied more than once the corresponding factor is $1/M^{L+k-1}$ at leading order in M. However, there are small corrections proportional to $1/M$. The presence of these corrections can be also understood noticing that for $M = 1$ the probability of a projected fat diagram with multiple occupancies must be zero. The above argument can be easily understood looking at Fig. 7.5 where the simplest case of a fat diagram with zero loops $L = 0$ and $k = 2$ external vertices is illustrated. In this case the fat diagram is just a line. In the leftmost part of Fig. 7.5 we show how, according to a probabilistic argument, in the case of the simple line we have a contribution $1/M$ to proceed from the beginning to the end of the line. Calling x_i the positions of the points on the line on the original lattice, $i = 0, ..., L$ and starting from a given layer α, one can choose one among the M layers for all the internal points. Once the summation over all possible realizations is performed, we obtain a factor 1. At the end, there is a link between x_{L-1} and x_L^β at fixed β with probability $\frac{1}{M}$.

On the right of Fig. 7.5 we see that the central point is occupied more than once. Naturally multiple occupations of the central point must correspond to different layers, otherwise the original fat diagram on the M-lattice would have a loop. Therefore, the graph on the right contributes as $\frac{1}{M}\left(1 - \frac{1}{M}\right)$, where the factor $\left(1 - \frac{1}{M}\right)$ is counting

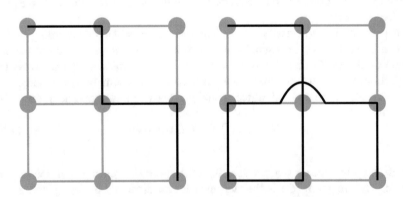

Fig. 7.5 The M dependence of the weight W for a fat diagram (black lines) is different if it has or not multiple occupancy points. For the left diagram we have $W = 1/M$ exactly, while for the right part we have $W = \frac{1}{M}\left(1 - \frac{1}{M}\right)$. The effect is sub-leading in powers of $1/M$, irrelevant when considering the critical behavior

the probability of choosing two different layers for the two occupations of the central point. The probability of this diagram is zero if $M = 1$ and it is smaller than the case where no multiple occupancies occur.

According to the argument above, one can conclude that: for a general fat diagram associated with a k-point observable (i.e. with k external legs) and L loops, at leading order we have the following simple expression:

$$W(G) \approx \frac{1}{M^{L+k-1}} . \tag{7.66}$$

We remind that the exact expression for $W(G)$ at all orders in M would include complicated $1/M$ corrections that are irrelevant as long as we are interested in the critical behavior.

7.6.4 Critical Behavior

The results of the previous sections are completely general and valid for any lattice. As mainly interested in the critical phenomena formalism, we want to study the expansion in the thermodynamic limit close to a phase transition. In the thermodynamic limit by definition the original lattice is infinite and we have to take into account an infinite sum over projected fat diagrams. Furthermore, we work under the assumption that the generic observable $O(G)$ (and thus $O_{lc}(G)$) is homogeneous. More precisely its value depends only on the topology of the graph G and on the length of its internal lines but not on the actual way through which the (projected) fat diagram is actually realized on the original lattice.

It is thus natural to order the sum (7.65) according to the topology of the projected fat diagram, i.e. in terms of Feynman diagrams. Since at leading order all diagrams with the same topology have the same $W(G) \approx 1/M^{L+k-1}$, we can consider their sum at fixed internal distances and ignore possible multiple occupancies that would lead to sub-leading terms.

This implies that the contribution of a given Feynman diagram at fixed lengths can be written as a sum, over the position of the internal vertices and the orientations of the lines entering into each vertex, of the product of the number of non-backtracking paths of the corresponding length for each line. For instance, with reference to diagram shown in Fig. 7.4, the sum of all projected fat diagrams with internal lines L_1, L_2, L_3 and L_4 contributes to the average of a two-point observable as:

$$\frac{1}{M^2 S(G)} \sum_{x_2,x_3,\mu_1,\mu_4,\{\mu_2,\nu_2\eta_2\},\{\mu_3,\nu_3\eta_3\}} \#_{L_1}(x_1, x_2, \mu_1, \mu_2) \#_{L_2}(x_2, x_3, \nu_2, \nu_3)$$

$$\cdot \#_{L_3}(x_2, x_3, \eta_2, \eta_3) \#_{L_4}(x_3, x_4, \mu_3, \mu_4) O_{lc}(L_1, L_2, L_3, L_4) .$$

In the expression above, x_1 and x_4 identify the positions of external vertices while x_2 and x_3 those of internal vertices in the fat diagram. $S(G)$ takes into account possible symmetries of the diagram which keeping its structure unchanged: it refers to the summation over both the positions of the internal vertices and the orientations of the lines meeting in each vertex. The role of the symmetry factor will be discussed in more details in Appendix D, but we want to remark here that is an intrinsic property of the Feynman diagram, being the same used in the field theoretical loop expansions.

The function $B_L(x, y, \mu, \nu)$ is the number of non-backtracking paths of length L from position x to y entering respectively from directions μ and ν. A detailed analysis of the function $B_L(x, y, \mu, \nu)$ is given in the appendix. If we assume that the number of paths does not depend on the directions (which is correct at large distances), we can write the contribution of the diagram as a sum over the position of the internal vertices of the product of a term $B_L(x_i, x_j)$ for each line (ij). The sum over the choice of the internal directions at each k-vertex provides a factor $\binom{2D}{k}k!$ which accounts for the number of ordered k-uples out of the $2D$ possible directions.

As before, with reference to the same fat diagram shown in the following, we have:

$$
\begin{aligned}
&\equiv \quad \frac{1}{M^2 S(G)} \sum_{x_2, x_3} (2D)^2 \left(\frac{(2D)!}{(2D-3)!} \right)^2 \#_{L_1}(x_1, x_2)\#_{L_2}(x_2, x_3) \\
&\quad \#_{L_3}(x_2, x_3)\#_{L_4}(x_3, x_4)O_{lc}(L_1, L_2, L_3, L_4)
\end{aligned}
$$

$$(7.67)$$

To determine the total contribution of a Feynman diagram we have to sum over the lengths of its lines. We will show that each Feynman diagram gives a contribution that diverges at the critical point and the divergence is determined by the behavior of the corresponding fat diagram when all lines are very large. The results obtained so far depend only on the M-layer construction. To go beyond, we need to specify the large distance behavior of $O_{lc}(G)$. This is a property of the actual problem we are studying, e.g. a finite-temperature phase transition with or without quenched disorder, a zero-temperature phase transition or a percolation problem or any other process that has a critical point on the Bethe lattice.

We will give a more detailed explanation of the behavior of $O_{lc}(G)$ in the next section, showing that its behavior at large distances generically determines the result. The simplest possibility occurs when *all* lengths are large and $O_{lc}(G)$ can be written as the sum of products over each line L_{ij} of the fat diagram of a factor $\lambda^{L_{ij}}$ with $\lambda \to \lambda_c = 1/(2D-1)$, when approaching the critical point. Again, with reference to the diagram above we would have:

$$O_{lc}(L_1, L_2, L_3, L_4) \approx c(G)\lambda^{L_1+L_2+L_3+L_4} ,\qquad (7.68)$$

where the unspecified constant $c(G)$ depends on the model and turns out to be essential in the computation of the critical exponents. The above expression is only valid at large distances but it gives the correct critical behavior. Indeed, if we insert it into Eq. (7.67) and sum over all internal lengths L_1, L_2, L_3 and L_4, we obtain an expression that depends on the generating function of the non-backtracking paths. If we move to Fourier space we arrive to an expression involving the Fourier transform $\hat{B}_\lambda(k)$ that has a singularity in $\lambda = \lambda_c$ and can be approximated at small momenta as $\hat{B}_\lambda^{-1}(k) \propto \lambda - \lambda_c + k^2$. For this analysis we refer the reader to Appendix C.

With reference again to the usual diagram in Eq. (7.67), its contribution turns out to be:

$$\frac{u^2}{M^2 S(G)} \frac{1}{m^2 + k^2} \left(\int d^D q \, \frac{1}{m^2 + q^2} \frac{1}{m^2 + (k-q)^2} \right) \frac{1}{m^2 + k^2} \qquad (7.69)$$

where $m^2 \equiv \lambda - \lambda_c$ and u is an effective (bare) coupling constant absorbing all vertex contributions and possible rescalings.

We have now all the tools to summarize the Feynman rules to determine the critical behavior of the $1/M$ expansion. Given a connected observable of order k, we have to sum over all connected Feynman diagrams with k external legs. For each Feynman diagram one has to:

- multiply a factor $1/M^{k-1+L}$ where L is the number of loops of the diagram;
- divide by the symmetry factor $S(G)$ of the graph;
- for each vertex of degree m multiply a factor $(2D)!/(2D - m)!$;
- study $O_{lc}(G)$ where G is the fat diagram with the topology of the Feynman diagram when its internal lines are large. This is the regime where the properties of the actual model enter into place. In the simplest case discussed above we would have a factor λ^L for each internal line and, as we saw before, we should then multiply by a factor $1/(m^2 + k^2)$ for each leg of the Feynman diagram. In general, we could have a more complicated situation. For instance, the contribution of a given line could be not just λ^L but rather $d^p/d\lambda^p \lambda^L$, which requires to multiply by a factor:

$$\frac{d^p}{d\lambda^p} \frac{1}{\lambda - \lambda_c + k^2} \qquad (7.70)$$

for the corresponding line.[10]
- perform the integration over the L momenta that are not fixed by momentum conservation at the vertices.

It should be clear that the analysis is only correct near a critical point and does not give the exact $1/M$ expansion but only the leading divergent part. The degree of divergence is expected to increase with the number of loops, and therefore at a given order in $1/M^p$ the maximally divergent part comes from the diagrams with $L = p + 1 - k$ loops.

---

[10]This phenomenon with $p = 1$ occurs in the random-field Ising model.

7.6.5   The Expression for Line-Connected Observables

We implicitly defined above the notion of line-connected observable through the following formula:

$$O(G) = \sum_{G' \subseteq G} O_{lc}(G') .$$

(7.71)

which can be written in the inverted notation as:

$$O_{lc}(G') = \sum_{G'' \subseteq G'} c_{G',G''} O(G'')$$

(7.72)

with appropriate coefficients $c_{G,G'}$.[11]

To better explain Eq. (7.71), we give a brief example of how computing a generic observable associated with the following two end-points one-loop diagram:

$$O\left(\!-\!\!\bigcirc\!\!-\!\right) = O_{lc}\left(\!-\!\!\bigcirc\!\!-\!\right) + O_{lc}\left(\!-\!\!\bigcirc\ \cdot\right) + O_{lc}\left(\cdot\ \bigcirc\!\!-\!\right) +$$
$$+ O_{lc}\left(\!-\!\!\frown\!\!-\!\right) + O_{lc}\left(\!-\!\!\smile\!\!-\!\right) + O_{lc}(\cdot\ \cdot) .$$

(7.73)

Applying again Eq. (7.71) to the *tadpole*, we get:

$$O\left(\!-\!\!\bigcirc\ \cdot\right) = O_{lc}\left(\!-\!\!\bigcirc\ \cdot\right) + O_{lc}(\cdot\ \cdot) .$$

(7.74)

Collecting together the expressions above and the following identity condition:

$$O(\cdot\ \cdot) = O_{lc}(\cdot\ \cdot) ,$$

(7.75)

we easily recover a compact expression for the line-connected observable, that is:

$$O_{lc}\left(\!-\!\!\bigcirc\!\!-\!\right) = O\left(\!-\!\!\bigcirc\!\!-\!\right) - O\left(\!-\!\!\frown\!\!-\!\right) - O\left(\!-\!\!\smile\!\!-\!\right) +$$
$$- O\left(\!-\!\!\bigcirc\ \cdot\right) - O\left(\cdot\ \bigcirc\!\!-\!\right) + 3O(\cdot\ \cdot) .$$

(7.76)

We immediately understand that if the observable O is connected, the last three terms, associated with disconnected diagrams, vanish. The diagram shown above contributes to the order $1/M^2$, as it corresponds to a single loop. The definition of line-connected observables, by subtracting the second and the third terms, allows

---

[11]They coefficients are also discussed in the literature of the so-called Moebius inversions in incidence algebra.

us to remove the contributions already considered in the $1/M$ term, i.e. single-line contributions considered as independent.

For our discussion it is nevertheless much more useful the following explicit formula:

$$O_{lc}(G) = \sum_{\tilde{G}' \subseteq G} (-1)^{N(G)-N(\tilde{G}')} O(\tilde{G}') ,$$ (7.77)

where $N(G)$ is the sum of all lines in graph G. Note that, compared to Eq. (7.72) that is defined on (sub)-fat-diagrams, here the sum runs in the class of tilded sub-diagrams $\tilde{G}$ including also graphs with dangling edges. The validity of the above explicit expression can be proved through the following steps. First we note that the above expression vanishes on a graph with at least one dangling edge:

$$O_{lc}(\tilde{G}) \equiv \sum_{\tilde{G}' \subseteq \tilde{G}} (-1)^{N(\tilde{G})-N(\tilde{G}')} O(\tilde{G}') = 0 , \qquad \text{if } \tilde{G} \neq G .$$ (7.78)

This is due to the fact that $O(\tilde{G})$ is equal to $O(G)$, where G is the graph obtained from $\tilde{G}$ removing all dangling edges. Therefore we can write:

$$O(G) = \sum_{G' \subseteq G} O_{lc}(G') = \sum_{\tilde{G}' \subseteq G} O_{lc}(\tilde{G}')$$ (7.79)

and then we can use the following identity to prove the formula for the line-connected observable:

$$\sum_{\{\tilde{G}' : \tilde{G}'' \subseteq \tilde{G}' \subseteq \tilde{G}\}} (-1)^{N(\tilde{G}')-N(\tilde{G}'')} = \delta(\tilde{G}, \tilde{G}'')$$ (7.80)

where $\delta(\tilde{G}, \tilde{G}'')$ is equal to one if $\tilde{G} = \tilde{G}''$ and zero otherwise. Note that the sum over all tilded $\tilde{G}'$, which are equal to a fat diagram G' once their dangling edges are removed, gives the factor $c_{G,G'}$ in the Moebius inversion formula [18].

The interpretation of the explicit formula (7.77) is straightforward: in order to compute $O_{lc}(G)$ we have to evaluate the observable O on each of the subgraphs that are obtained from G by removing sequentially its lines times a factor -1 for each line removed.

Given these premises, we can now discuss the main difference between $O(G)$ and $O_{lc}(G)$. It lies in their different behavior when one of the lines in G, say l', tends to infinity. In this case, $O(G)$ tends to a constant: the value that it has on the subgraph where l' has been removed. On the other hand $O_{lc}(G)$ can be written as the sum of certain diagrams with l' minus the same expression evaluated on diagrams without l'. Since the first expression tends to the second one when l' tends to infinity, $O_{lc}(G)$ itself tends to zero. This is also why we use the name *line-connected* observables.

In this spirit, we recall that in the previous section we ordered the sum over fat diagrams collecting all of them with the same topology and summed over all possible internal lengths. If $O_{lc}(G)$ tends to a constant, when one of its internal lines tends to infinity, the procedure is unavoidably in trouble because the contribution of the Feynman diagram would be also infinite.

To conclude this section we want to give an example taken from statistical mechanics, that is to consider a lattice whose nodes are occupied by real variables σ_i. The most general two-point observable is the probability $P(\sigma, \tau)$ defined according to the Gibbs distribution of the problem. In this case the computation of the $O(G)$ amounts to: (i) multiply a term $P_L(\sigma_i, \sigma_j)$ for each line L_{ij} of length L; (ii) multiply a term $P^{2D-k}(\sigma)$ for each vertex of degree k; (iii) sum over the configuration of the internal vertices. $P_L(\sigma_i, \sigma_j)$ is the probability distribution on a one-dimensional line where the two end-points have degree one and $P(\sigma)$ is the solution of the iterative equation. With reference to the diagram discussed above in Eq. (7.67), we have:

$$O(G) = \frac{1}{\mathcal{N}(G)} \sum_{\sigma_2, \sigma_3} P^{2D-1}(\sigma_1) P_{L_1}(\sigma_1, \sigma_2) P^{2D-3}(\sigma_2) P_{L_2}(\sigma_2, \sigma_3) P_{L_3}(\sigma_2, \sigma_3) \cdot$$
$$\cdot \ P^{2D-3}(\sigma_3) P_{L_4}(\sigma_3, \sigma_4) P^{2D-1}(\sigma_4)$$

$$(7.81)$$

where the normalization $\mathcal{N}$ reads:

$$\mathcal{N}(G) = \sum_{\sigma_1, \sigma_2, \sigma_3, \sigma_4} P^{2D-1}(\sigma_1) P_{L_1}(\sigma_1, \sigma_2) P^{2D-3}(\sigma_2) P_{L_2}(\sigma_2, \sigma_3) P_{L_3}(\sigma_2, \sigma_3) \cdot$$
$$\cdot \ P^{2D-3}(\sigma_3) P_{L_4}(\sigma_3, \sigma_4) P^{2D-1}(\sigma_4) \ .$$

$$(7.82)$$

To compute $O_{lc}(G)$ we have to remove sequentially all the lines in G. Computing O on the diagram without a line L_k amounts to evaluate the above expression where we replace:

$$P_{L_k}(\sigma_i, \sigma_j) \rightarrow P(\sigma_i) P(\sigma_j) \ . \tag{7.83}$$

As we saw in the previous section, we are interested in the behavior of $O_{lc}(G)$ when all lines are large. This is controlled by the behavior of $P_L(\sigma, \tau)$ for large L. Let us assume that we are in the simplest case in which we have in the large-L limit:

$$P_L(\sigma, \tau) \approx P(\sigma) P(\tau) + \lambda^L a(\sigma) a(\tau) \ , \tag{7.84}$$

where λ is the critical eigenvalue equal to $\lambda_c = 1/(2D - 1)$ precisely at the critical point. We can now see that the denominator of all tilded subdiagrams $\tilde{G} \in G$ is given at leading order by:

$$\mathcal{N}(\tilde{G}) \approx \left(\sum_\sigma P^{2D}(\sigma) \right)^4 , \tag{7.85}$$

where the exponent four comes from the number of vertices. Since the denominator of $O(\tilde{G})$ is the same at leading order, we can now collect all numerators of the diagrams involved in $O_{lc}(G)$. Now the peculiar form of Eq. (7.77) takes place: since it requires to sum over all lines with a factor -1 for each removed line, we can write:

$$O_{lc}(G) \approx \frac{1}{\left(\sum_\sigma P^{2D}(\sigma)\right)^4} \sum_{\sigma_2,\sigma_3} P^{2D-1}(\sigma_1)\Delta P_{L_1}(\sigma_1,\sigma_2) P^{2D-3}(\sigma_2)\Delta P_{L_2}(\sigma_2,\sigma_3)$$

$$\Delta P_{L_3}(\sigma_2,\sigma_3) P^{2D-3}(\sigma_3)\Delta P_{L_4}(\sigma_3,\sigma_4) P^{2D-1}(\sigma_4) \,,$$

where

$$\Delta P_L(\sigma,\tau) \equiv P_L(\sigma,\tau) - P(\sigma)P(\tau) \approx \lambda^L a(\sigma)a(\tau) \,. \tag{7.86}$$

This implies:

$$O_{lc}(G) \approx \frac{\left(P^{2D-1}(\sigma_1)a(\sigma_1)\right)\left(P^{2D-1}(\sigma_4)a(\sigma_4)\right)\left(\sum_\sigma P^{2D-3}(\sigma)a^3(\sigma)\right)^2}{\left(\sum_\sigma P^{2D}(\sigma)\right)^4}\lambda^{L_1+L_2+L_3+L_4}$$

$$\tag{7.87}$$

that can be easily generalized to higher order connected correlations and more Feynman diagrams involving vertices of all degrees. For this specific problem, once we have summed over the length of all lines, we should multiply by a factor:

$$\frac{\left(P^{2D-k}(\sigma)a^k(\sigma)\right)}{\left(\sum_\sigma P^{2D}(\sigma)\right)} \quad \text{and} \quad \frac{\left(\sum_\sigma P^{2D-k}(\sigma)a^k(\sigma)\right)}{\left(\sum_\sigma P^{2D}(\sigma)\right)} \tag{7.88}$$

for each external and internal vertex of degree k respectively. One should consider also a factor $1/(\lambda - \lambda_c + k^2)$ for each line. This holds for all Feynman diagrams of all connected correlations and, since the symmetry factors are the same, we safely conclude that the critical behavior of such a theory is completely equivalent to a scalar cubic theory, as required by universality.

In general, for any phase transition in the universality class of a given field theory one should be able (maybe with some exceptions) to recover the very same field theory from the $1/M$ expansion. Much more interesting are the cases in which no field-theoretical mapping is known, including notably disordered systems at zero temperature and various percolation problems. In these more complex cases one should evaluate $O_{lc}(G)$ explicitly. If O is a connected observable, one should only consider connected subgraphs. For instance, with reference to the usual diagram above, one should sum O evaluated on the original diagram and subtract its value on the diagrams with lines L_2 and L_3 missing.

It is also instructive to check how the above mapping to a scalar ϕ^3 theory behaves for a ferromagnetic transition, in which we expect instead a mapping to a ϕ^4 theory. The key point is the internal vertex:

$$V \equiv \frac{\left(\sum_\sigma P^{2D-3}(\sigma)a^3(\sigma)\right)}{\left(\sum_\sigma P^{2D}(\sigma)\right)} \,. \tag{7.89}$$

It turns out that in the case of a ferromagnetic transition $P(\sigma)$ is even as a function of σ, while the critical eigenvector $a(\sigma)$ is odd. Therefore, the prefactor of the cubic vertex is zero. To get a finite contribution we cannot attach three terms $\lambda^L a(\sigma)a(\tau)$ to a cubic vertex but we have to consider the less divergent terms in the expansion (7.84).

This implies that at least one of the legs of the vertex must give a non-divergent contribution and this acts as a contraction of two vertices of degree three into a single vertex of degree four.[12] In the second diagram in Fig. 7.4 we have to assume that one of the two internal legs remains finite. Analogously, in the first diagram of the same series, the line connecting the two internal vertices must also remain finite. The two diagrams are thus equivalent to that with one loop and one quartic vertex that provides the first correction to the self-energy in an ordinary quartic theory.

7.7   Conclusions

Starting from a fully-connected ferromagnetic model, the Curie-Weiss model, we generalized our discussion to an auxiliary model defined by replicating M-times the original lattice. On it we computed $1/M$ corrections to the leading order Bethe solution and we showed that this expansion corresponds to a loop expansion, where now the loops have to be meant as spatial ones. Both the critical properties and the symmetry factors of the diagrams coincide in the two approaches and depend only on the choice of topology. Using a cavity argument we introduced a topological expansion in terms of *fat diagrams*, valid for both ordered and disordered systems. Essentially a fat diagram is a graph with k external vertices, no dangling edges and no completely disconnected components.

To define a suitable expansion in terms of such fat diagrams, we introduced the notion of *line-connected* observable on the M-layer, including also contributions from sub-diagrams. This observable differs from a non-line-connected one for its asymptotic behavior when the internal lines go to infinity. According to this supplementary definition, we managed to write a perturbative series of all fat diagrams and achieve deeper insights of the critical properties of the model.

References

1.   Abou-Chacra R, Thouless DJ, Anderson PW (1973) A selfconsistent theory of localization. J Phys C Solid State Phys 6(10):1734
2.   Altieri A, Parisi G, Rizzo T (2016) Composite operators in cubic field theories and link-overlap fluctuations in spin-glass models. Phys Rev B 93(2):024422
3.   Altieri A et al (2017) Loop expansion around the Bethe approximation through the M-layer construction. J Stat Mech: Theory Exp 2017(11):113303
4.   Baity-Jesi M et al (2014) Dynamical transition in the D= 3 Edwards-Anderson spin glass in an external magnetic field. Phys Rev E 89(3):032140
5.   Baity-Jesi M et al (2014) The three-dimensional Ising spin glass in an external magnetic field: the role of the silent majority. J Stat Mech: Theory Exp 2014(5):P05014

---

[12]This framework allowed us to clarify the reason why the diagram shown in (7.67) is not permitted in a ferromagnetic model, where it would have instead the properties of a tadpole.

6. Biroli G, Semerjian G, Tarzia M (2010) Anderson model on Bethe lattices: density of states, localization properties and isolated eigenvalue. Prog Theor Phys Suppl 184:187
7. Brezin E, Zinn-Justin J, Le Guillou JC (1976) Phase transitions and critical phenomena, vol 6. Ed. by Domb C, Green MS. Academic Press
8. Cammarota C et al (2013) Fragility of the mean-field scenario of structural glasses for disordered spin models in finite dimensions. Phys Rev B 87(6):064202
9. Chertkov M, Chernyak VY (2006) Loop calculus in statistical physics and information science. Phys Rev E 73(6):065102(R)
10. De Dominicis C, Giardina I (2006) Random fields and spin glasses. Cambridge University Press
11. Efetov KB (1990) Effective medium approximation in the localization theory: saddle point in a lagrangian formulation. Phys A Stat Mech Appl 167(1):119
12. Ferrari U et al (2013) Finite-size corrections to disordered systems on Erdøs-Rényi random graphs. Phys Rev B 88(18):184201
13. Fitzner R, van der Hofstad R (2013) Non-backtracking random walk. J Stat Phys 150(2):264
14. Itzykson C, Drouffe J-M (1991) Statistical field theory: volume 2, strong coupling, Monte Carlo methods, conformal field theory and random systems, vol 2. Cambridge University Press
15. Itzykson C, Zuber JB (1980) Quantum field theory. McGraw-Hill
16. Janson S, Luczak T, Rucinski A (2011) Random graphs, vol 45. Wiley
17. Kschischang FR, Frey BJ, Loeliger H-A et al (2001) Factor graphs and the sum-product algorithm. IEEE Trans Inf Theory 47(2):498
18. Kung JPS, Rota G-C, Yan CH (2009) Combinatorics: the Rota way. Cambridge University Press
19. Le Bellac M (1991) Quantum and statistical field theory. Clarendon Press
20. Lucibello C et al (2014) Finite-size corrections to disordered Ising models on random regular graphs. Phys Rev E 90(1):012146
21. Montanari A, Mézard M (2009) Information, physics and computation. Oxford University Press
22. Montanari A, Rizzo T (2005) How to compute loop corrections to the Bethe approximation. J Stat Mech: Theory Exp 2005:P10011
23. Mori R (Mar 2013) New understanding of the Bethe approximation and the replica method. PhD thesis. University of Kyoto, p 3
24. Mori R, Tanaka T (Oct 2012) New generalizations of the Bethe approximation via asymptotic expansion. In: Proceedings of IEICE SITA
25. Parisi G (1988) Statistical field theory. Addison-Wesley
26. Parisi G, Mézard M, Virasoro MA (1987) Spin glass theory and beyond. World Scientific Singapore
27. Parisi G, Ricci-Tersenghi F, Rizzo T (2014) Diluted mean-field spin-glass models at criticality. J Stat Mech: Theory Exp 2014(4):P04013
28. Parisi G, Slanina F (2006) Loop expansion around the Bethe-Peierls approximation for lattice models. J Stat Mech: Theory Exp 2006(02):L02003
29. Parisi G, Sourlas N (1979) Random magnetic fields, supersymmetry, and negative dimensions. Phys Rev Lett 43(11):744
30. Raymond J, Ricci-Tersenghi F (2017) Improving variational methods via pairwise linear response identities. J Mach Learn Res 18(6):1
31. Rizzo T, Wemmenhove B, Kappen HJ (2007) Cavity approximation for graphical models. Phys Rev E 76(1):011102
32. Sacksteder VE (2007) Sums over geometries and improvements on the mean field approximation. Phys Rev D 76(10):105032
33. Vontobel PO (2013) Counting in graph covers: a combinatorial characterization of the Bethe entropy function. IEEE Trans Inf Theory 59(9):6018
34. Yedidia JS, Freeman WT, Weiss Y (2003) Understanding belief propagation and its generalizations. Explor Artif Intell New Millenn 8:236
35. Zinn-Justin J (2002) Quantum field theory and critical phenomena. Oxford Science Publications

Part III
Conclusions

Chapter 8
Conclusions and Perspectives

This thesis has been devoted to the investigation of several aspects of interest in disordered systems, with both quenched or self-generated disorder. The aim was actually twofold. In the first part, a special emphasis has been put on the study of the perceptron model in the non-convex phase, where new unexpected properties emerge. In this regime, it has been shown that the perceptron falls in the same universality class as hard spheres in infinite dimensions. Our approach is based on a perturbative expansion of the free energy in the SAT (hard-sphere) regime of the perceptron phase diagram. Although we retained only the first two moments of the expansion—an approximation motivated by the fully-connected structure of the model—significant results have been obtained in the perceptron and related fields. Because considering fully-connected models might sometimes be reductive or completely misleading, the second part focuses on the Bethe multi-layer construction, which lays the groundwork to study finite-size corrections in lattice models.

As each of the previous chapters already contained a conclusive remark section, in the following we will summarize our main results and mention possible directions for further research.

8.1 Summary of the Main Results

Jamming transition and applications

- In Chap. 4 we proposed the derivation of an effective thermodynamic potential (Sect. 4.4), valid in both the high-temperature and in the low-temperature regimes. Our method also led to an exact derivation in high dimensions of a logarithmic interaction dominating the hard-sphere regime close to the jamming line (Sect. 4.5). We confirmed and generalized the argument proposed in [6] for three-dimensional hard-sphere glass formers. Moreover, we exactly derived the spectrum of low-

© Springer Nature Switzerland AG 2019
A. Altieri, *Jamming and Glass Transitions*, Springer Theses,
https://doi.org/10.1007/978-3-030-23600-7_8

energy excitations, in both the liquid phase and the jamming limit. In particular, in the SAT phase, we highlighted the divergent behavior of the spectral density, governed by a non-trivial exponent above the cut-off frequency (Sect. 4.8.3). This exponent, which is in remarkable agreement with the results of Wyart and coworkers [11], can be also related to the critical exponent that governs the force distribution at jamming.

- From the analysis of the effective potential near the jamming line, we also derived the leading and sub-leading contributions in the inter-particle forces (Sect. 4.7). In this context, the discussion about peculiar scaling laws dominating the jamming phase naturally led to the investigation of a crossover regime between the SAT and the UNSAT phases. The derivation of a crossover temperature played a central role as it allowed to define two different regimes: below that, the system behaves like a zero-temperature assembly of soft particles, otherwise it enters the entropic-like regime.

- A further detailed analysis concerned third and higher-order corrections to the effective potential that turn out to be relevant either in accounting for finite-dimensional systems, not exactly at jamming, or in numerical simulations. Surprisingly, we showed that higher-order corrections to the effective potential, except for the first two moments, do not contribute in the jamming limit (Sect. 4.6). This result remarks once more the connection between jamming and a mean-field-like scenario.

- Keeping in mind the connection between the perceptron and sphere packings in high dimensions, we extended our predictions to derive an analogous effective potential in sphere models, very beneficial to describe critical phases as well. To bridge the gap with the TAP approach, we rephrased the computations in the replica formalism (Sect. 5.3). In any case, introducing the TAP free energy has the clear advantage to provide a more compact and straightforward framework.

  The study of hard-sphere models posed new intriguing questions, such as the possibility (or the impossibility) to observe a phenomenon similar to the Bose-Einstein condensation in real space, or even to analyze critical properties of economic networks and ecosystems. The common thread emerging here is the complex free-energy landscape occurring in high-dimensional settings which, in terms of stability and improvement of the network, appears really different from low-dimensional systems.

- Connected to ecosystem problems, one of the main goal of this thesis was to highlight new links between the perceptron model and a recent high-dimensional variant of ecological feedback (Sect. 6.3). We studied the analytical background behind it to connect its main features to the statistical physics of glasses. We especially emphasized the importance of emerging zero modes, possibly related to the definition of diverging susceptibilities, and the analogy between a cohesion effect in a multi-species community and an isostatic regime (Sects. 6.4–6.5).

Field-Theory Methods Beyond a Fully-Connected Topology
The message that should appear clear at this point concerns the depth of predictions that can be obtained in the framework of high-dimensional or fully-connected mod-

els. However, a naïve mean-field approach might be unable to describe finite-size corrections and interesting transitions occurring in finite-dimensional models or in finite-connectivity random graphs. One explication for this deficiency is that a loop expansion around the fully-connected solution does not account correctly for the emergence of multiple solutions and then for loopy structures.

Therefore, we proposed in Chap. 7 the Bethe M-layer construction allowing to interpolate between the large M limit, for which the Bethe approximation is exact, and the original model, for $M = 1$. We defined a $1/M$ perturbative series corresponding to a diagrammatic expansion in terms of the so-called fat diagrams (Sect. 7.6.2). The central motivation for this study was actually linked to the possibility to have a systematic expansion for whatever problem studied on a Bethe lattice. This includes, of course, statistical mechanics models at finite temperature, with or without quenched disorder, but also systems for which we cannot define a Hamiltonian at all, such as some instances of percolation problems or zero-temperature systems.

8.2 Perspectives and Future Developments

Our work enriches the analysis of the physics of disordered systems and the jamming transition, focusing in particular on a systematic free-energy expansion, developed in the perceptron and high-dimensional sphere models, and on the Bethe M-layer construction. In view of the results achieved thus far, we would like to draw further perspectives that could be of interest in the very next steps of the study of glassy systems at large.

8.2.1 Perceptron Model

The possibility to obtain exact results in the perceptron model is one of the compelling reasons for studying it. First, we would like to have a better numerical control of the boundary layer in the perceptron phase diagram between the SAT and the UNSAT phases. In Chap. 4 we discussed how to derive a crossover temperature T^* that can be directly related to the pressure and other physical parameters of the model. It signals the crossover between a zero-temperature soft-sphere regime and an anharmonic one. In [10] scaling solutions have been proposed to describe the emergence of anharmonic effects at finite temperature. A possible way to investigate the different regimes to compare with [10] would be to numerically study the full RSB solution across the transition lines and to reproduce and extend some of the results of [12].

Second, studying a random dilute version of the perceptron might be of great interest for the glass literature. A dilute perceptron could be defined by forcing the pattern connectivity to be finite or by constraining the single dynamical particle to interact only with a certain set of obstacles. An analysis of this type can be done via the *cavity method*. In principle, this would allow us to generalize our predictions beyond mean field and to understand whether different universality classes exist. This working scenario can be somehow related to the computation of higher-order corrections to the effective potential presented in Sect. 4.6 in order to identify the true (mean field or not) nature of the jamming transition.

Furthermore, the vibrational modes we obtained from the Marcenko-Pastur analysis are purely delocalized. On the other hand, in low dimensions some modes are quasi-localized. An open question is to find the microscopic origin of these modes. We could attempt to use our expansion, originally conceived around the $d \to \infty$ solution, to make predictions and make sense of this process.

8.2.2   Hard-Sphere Models

The picture that emerges from this work suggests the absence of a condensation transition in real space for high-dimensional hard-sphere models. The weakness of our approach relies on the lack of a notion of distance. Therefore, instead of considering a hyper-spherical manifold, we could re-formulate the problem on a graph by fixing the particle coordination number c. The balance between the entropic and the energetic contributions in the free energy could be then evaluated numerically. We would introduce in this way a topology and bridge the gap with finite-dimensional systems, where this kind of transition is actually observed. There should exist a critical density in the interaction network, above which a crossover between a *fluid* normal phase and a mixed *condensed* phase should occur. The interest behind this issue comes from the fact that several instances of complex systems—such as the World Wide Web, stochastic models of meta-community ecosystems, business world companies—are characterized by a peculiar topology where nodes tend to self-organize into complex clusters. The fact that these models are fairly well reproduced by a Bose-Einstein statistics—where nodes correspond to energy levels and links to particles—and eventually exhibit a transition, for which a single node captures a finite fraction of links, has obviously important applications in physics and beyond [1, 2, 5].

From an analytical perspective, an alternative approach might be to introduce in the model an auxiliary tunable parameter, similarly to the construction proposed in Chap. 7, which would allow us to interpolate between the original highly-connected version and a finite model. In a sense, this construction is reminiscent of the original Mari-Kurchan model.

To give some hints, we might be interested in a system that has a high, but finite, connectivity locally and that can be mapped to finite dimensions. We might consider the model in $N + d_f$ dimensions, where the first N coordinates are real physical dimensions, which extend to infinity in the thermodynamic limit, while the others d_f are auxiliary dimensions, that can be either wrapped together or not, but are supposed to remain finite. However, if one considers the limit $d_f \to \infty$, the solution locally converges to the infinite-dimensional one. Playing with d_f and progressively increasing its value, the system can be mapped to a fully-connected model, but for each d_f it keeps its finite-size structure. The different nature of the particles, namely their poly-dispersity, should play a fundamental role in any case. This might be a reasonable and testable starting point, as several models that undergo a non-trivial phase transition, favoring the formation of a mixed *condensed* phase, are defined on a lattice.

8.2.3 Connections Between Jamming and Ecology

At the interface between different domains, the jamming paradigm might be interesting not only in relation to optmization problems and sphere packings but even to ecology. At first glance two main aspects emerge when one studies these problems: (i) the network of connections between generic constraint satisfaction problems and biological/ecological settings; (ii) the possibility to extend our formalism beyond a convex-like version. It is well known that upon approaching the jamming line in the perceptron model the isostatic regime is associated with an abundance of zero modes. We might ask what the physical implications of an analogous phenomenon in ecology can be in terms of diverging susceptibilities.

Concerning the extension to a non-convex setting, the most natural thing might be to re-define the resource surplus via negative requirements, i.e. $\chi_\mu < 0$, which means that a given species plays a twofold role, requiring nutrients and producing a certain amount of a specific element. In this case, the parameter χ_μ should play the same role as σ for the negative perceptron model. Alternatively, one can impose an external physical constraint on the availabilities h_i to restrict the space of allowed solutions. This variant should be tested numerically to check whether a non-convex phase actually occurs. It can also lead to a more detailed study of the complex fitness landscapes.

It is worth noticing that the formalism used for perceptron, which is clearly a very abstract model, can be applied to a context of interacting agents, drawing not only a working strategy but also insights. To study the presence of a possible full RSB phase, we propose two possible routes. One has been mentioned above, with reference to the model by Tikhonov and Monasson, to be eventually analyzed in the non-convex regime. Alternatively, one could study the random replicant model [4] (which is very close to the perceptron but with an additional constraint on the squared moment of the concentrations of families in the system and characterized by a clear replica symmetry breaking solution) and its variants [3], map the respective phase diagrams and study the emergence of a full RSB at zero and finite temperatures.

Concerning resource optimality, an underlying assumption in many ecological models is that species communities get reasonably close to the optimal use of all the resource influxes, or fall on some Pareto surface if there are tradeoffs between various contributions. In line with [13], another possible attempt would be to propose a more robust and detailed explanation for these issues–relevant also for socio-economic systems [8] and metabolic networks [7, 9]—and to study more systematically the consistency of this hypothesis.

References

1. Bianconi G, Barabási A (2001) Bose-Einstein condensation in complex networks. Phys Rev Let 86(24):5632
2. Bianconi G, Ferretti L, Franz S (2009) Non-neutral theory of biodiversity. EPL (Europhysics Letters) 87(2):28001
3. Biroli G, Bunin G, Cammarota C (2018) Marginally stable equilibria in critical ecosystems. New J Phys 20(8):083051
4. Biscari P, Parisi G (1995) Replica symmetry breaking in the random replicant model. J Phys A: Math Gen 28(17):4697
5. Bouchaud J-P, Mézard M (2000) Wealth condensation in a simple model of economy. Phys A: Stat Mech Appl 282(3–4):536
6. Brito C, Wyart M (2009) Geometric interpretation of previtrification in hard sphere liquids. J Chem Phys 131(2):024504
7. De Martino A, Marinari E (2010) The solution space of metabolic networks: producibility, robustness and fluctuations. J Phys Conf Ser 233(1):012019. IOP Publishing
8. De Martino A, Marsili M (2006) Statistical mechanics of socio-economic systems with heterogeneous agents. J Phys A: Math Gen 39(43):R465
9. De Martino A, et al (2012) Reaction networks as systems for resource allocation: a variational principle for their non-equilibrium steady states. PloS One 7(7):e39849
10. DeGiuli E, Lerner E, Wyart M (2015) Theory of the jamming transition at finite temperature. J Chem Phys 142(16):164503
11. DeGiuli E, et al (2014) Force distribution affects vibrational properties in hardsphere glasses. Proc Natl Acad Sci 111(48):17054
12. Franz et al S (2017) Universality of the SAT-UNSAT (jamming) threshold in nonconvex continuous constraint satisfaction problems. SciPost Phys 2(3):019
13. Yoshino Y, Galla T, Tokita K (2007) Statistical mechanics and stability of a model eco-system. J Stat Mech: Theory Exp 2007(09):P09003

Appendix A
$O(\eta^3)$ Corrections to the Effective Potential in the Perceptron Model

In this Appendix we present a detailed derivation of the third-order term in the expansion of the Thouless-Anderson-Palmer (TAP) free energy. In principle, such corrections can be neglected both in the perceptron model considered thus far or in high-dimensional sphere systems, thanks to their fully-connected structure and to the thermodynamic limit to be performed at the end. However, one might be interested in defining a modified version of the perceptron on a finite-dimensional lattice to evaluate, for instance, finite-size corrections to the free energy. In that case, a detailed analysis beyond mean-field would be needed.

We remind in particular Eq. (4.59) as reported in the main text:

$$\Gamma(\vec{m}, \vec{f}) = \sum_i m_i u_i + \sum_\mu f_\mu v_\mu - \log \int d\vec{x} d\vec{h} d\hat{\vec{h}} \, e^{S_\eta(\vec{x}, \vec{h}, \hat{\vec{h}})} \qquad \text{(A.1)}$$

$$S_\eta(\vec{x}, \vec{h}, \hat{\vec{h}}) = \sum_i u_i x_i + \sum_\mu i v_\mu \hat{h}_\mu - \lambda \sum_i (x_i^2 - N) - \frac{\beta}{2} \sum_\mu h_\mu^2 \theta(-h_\mu) +$$
$$- i \sum_\mu \hat{h}_\mu (h_\mu - \eta h_\mu(x)) - \frac{b}{2} \sum_\mu (\hat{h}_\mu^2 - \alpha N \tilde{r}) .$$

$$\text{(A.2)}$$

where we introduced the auxiliary variables $i\hat{h}_\mu$, conjugated to the gaps, and the Lagrange multipliers u_i and v_μ, to enforce the average values of the forces and the positions between particles. Additional Lagrange multipliers, λ and b, accounts respectively for the correct normalization over the sphere and the constraint over the second moment of $i\hat{h}_\mu$.

© Springer Nature Switzerland AG 2019
A. Altieri, *Jamming and Glass Transitions*, Springer Theses,
https://doi.org/10.1007/978-3-030-23600-7

The first term in the small-coupling expansion corresponds to the average effective Hamiltonian, while the second term includes both the contribution of the connected correlation function for the Hamiltonian and the derivatives with respect to the associated Lagrange multipliers:

$$\frac{\partial^2 \Gamma}{\partial \eta^2} = -\left\{ \langle H_{\text{eff}}^2 \rangle - \langle H_{\text{eff}} \rangle^2 + \langle H_{\text{eff}} \left[\sum_i \frac{\partial u_i}{\partial \eta}(x_i - m_i) + \sum_\mu \frac{\partial v_\mu}{\partial \eta}(i\hat{h}_\mu - f_\mu) \right] \rangle \right\} =$$

$$= \alpha N (\tilde{r} - r)(1 - q) .$$

(A.3)

In the following we shall explain why off-diagonal terms, namely the ones with all different indices $(ij, \mu\nu)$, do not contribute to the potential. For the connected part in Eq. (A.3) we have:

$$\langle H_{\text{eff}}^2 \rangle - \langle H_{\text{eff}} \rangle^2 = \sum_{ij,\mu\nu} \frac{\xi_i^\mu \xi_j^\nu}{N} \langle x_i x_j i\hat{h}_\mu i\hat{h}_\nu \rangle_c .$$

(A.4)

The sum in Eq. (A.4) gets contributions from the following possible combinations of indices:

- $i \neq j, \mu \neq \nu$,
- $i = j, \mu \neq \nu$: $\left(\langle x_i^2 \rangle - m_i^2 \right) f_\mu f_\nu$,
- $i \neq j, \mu = \nu$: $\left(\langle (i\hat{h}_\mu)^2 \rangle - \langle i\hat{h}_\mu \rangle^2 \right) \langle x_i \rangle \langle x_j \rangle = \left(\langle (i\hat{h}_\mu)^2 \rangle - \langle i\hat{h}_\mu \rangle^2 \right) m_i m_j$,
- $i = j, \mu = \nu$: $\left(\langle x_i^2 \rangle \langle (i\hat{h}_\mu)^2 \rangle - \langle x_i \rangle^2 \langle i\hat{h}_\mu \rangle^2 \right)$.

In the first case, the contribution is simply zero according to the *clustering property* [2, 3]. We focus then on the last term in Eq. (A.3), which accounts for the derivatives of the additional Lagrange multipliers with respect to the parameter η:

$$\langle H_{\text{eff}} \left[\sum_j \frac{\partial u_j}{\partial \eta}(x_j - m_j) + \sum_\nu \frac{\partial v_\nu}{\partial \eta}(i\hat{h}_\nu - f_\nu) \right] \rangle =$$

$$= \langle \sum_{i,\mu} \frac{\xi_i^\mu x_i i\hat{h}_\mu}{\sqrt{N}} \sum_{j,\nu} \frac{\xi_j^\nu f_\nu}{\sqrt{N}}(x_j - m_j) \rangle + \langle \sum_{i,\mu} \frac{\xi_i^\mu x_i i\hat{h}_\mu}{\sqrt{N}} \sum_{j,\nu} \frac{\xi_j^\nu m_j}{\sqrt{N}}(i\hat{h}_\nu - f_\nu) \rangle =$$

$$= \frac{1}{N} \sum_{ij,\mu\nu} \left[\xi_i^\mu \xi_j^\nu f_\mu f_\nu (\langle x_i x_j \rangle) - m_i m_j) + \xi_i^\mu \xi_j^\nu m_i m_j (\langle i\hat{h}_\mu i\hat{h}_\nu \rangle - f_\mu f_\nu) \right] .$$

(A.5)

To be clearer in the derivation, we consider some examples that might confirm why only the terms with equal indices contribute to the effective potential. If we suppose to neglect for simplicity reasons the contribution coming from the patterns, we have then:

- $i = j, \mu \neq \nu$: $\left(\langle x_i^2 \rangle - m_i^2\right) f_\mu f_\nu$ that cancels with the second term in the list above
- $i \neq j, \mu = \nu$: $\left(\langle (i\hat{h}_\mu)^2 \rangle - \langle i\hat{h}_\mu \rangle^2\right) m_i m_j$ that cancels with the third term
- $i = j, \mu = \nu$: $\alpha r (1 - q) + \alpha q (\tilde{r} - r)$.

It is immediate to figure out that the only relevant contribution comes from the case $i = j$ & $\mu = \nu$, otherwise the correlation functions are zero either because they are totally disconnected (if $i \neq j, \mu \neq \nu$) or because they sum up to zero combined with another term of opposite sign.

Given these premises, we can evaluate third-order corrections, whose generic expression reads [1, 4]:

$$\frac{\partial^3 \Gamma}{\partial \eta^3} = \langle H_{\text{eff}} \rangle \frac{\partial \langle H_{\text{eff}} \rangle}{\partial \eta} + \langle H_{\text{eff}} \Upsilon_2 \rangle + \langle H_{\text{eff}} \left(H_{\text{eff}} - \langle H_{\text{eff}} \rangle + \Upsilon_1\right)^2 \rangle \qquad (A.6)$$

where Υ_n is:

$$\Upsilon_n = \sum_i \frac{\partial}{\partial y_i} \left(\frac{\partial^n \Gamma}{\partial \eta^n}\right) (s_i - y_i) . \qquad (A.7)$$

For simplicity, we indicate both derivatives, with respect to the particle positions and the forces, as $(s_i - y_i) \frac{\partial}{\partial y_i}$. This allows us to write the expression to the third order in a more compact way:

$$\frac{\partial^3 \Gamma}{\partial \eta^3} = - \langle H_{\text{eff}} \rangle \frac{\partial^2 \Gamma}{\partial \eta^2} + \langle H_{\text{eff}} \Upsilon_2 \rangle + \langle H_{\text{eff}} \left(H_{\text{eff}} - \langle H_{\text{eff}} \rangle + \Upsilon_1\right)^2 \rangle =$$

$$= \langle H_{\text{eff}} \rangle \left[\langle H_{\text{eff}}^2 \rangle - \langle H_{\text{eff}} \rangle^2 - \langle H_{\text{eff}} \sum_i (s_i - y_i) \frac{\partial \langle H_{\text{eff}} \rangle}{\partial y_i} \rangle \right] +$$

$$+ \langle H_{\text{eff}} \sum_i (s_i - y_i) \frac{\partial}{\partial y_i} \frac{\partial^2 \Gamma}{\partial \eta^2} \rangle + \langle H_{\text{eff}} \left(H_{\text{eff}} - \langle H_{\text{eff}} \rangle - \sum_i \frac{\partial \langle H_{\text{eff}} \rangle}{\partial y_i} (s_i - y_i) \right)^2 \rangle .$$

$$\qquad (A.8)$$

Expanding the last squared term and explicitly differentiating it with respect to m_i and f_μ, we obtain:

$$\frac{\partial^3 \Gamma}{\partial \eta^3} = \langle H_{\text{eff}}^3 \rangle + \langle H_{\text{eff}} \rangle \langle H_{\text{eff}}^2 \rangle - 2\langle H_{\text{eff}} \rangle^3 + \langle H_{eff} \rangle \langle H_{eff} \sum_i \frac{\partial \langle H_{\text{eff}} \rangle}{\partial m_i}(x_i - m_i)\rangle +$$

$$+\langle H_{\text{eff}} \rangle \langle H_{\text{eff}} \sum_\mu \frac{\partial \langle H_{\text{eff}} \rangle}{\partial f_\mu}(i\hat{h}_\mu - f_\mu)\rangle + \langle H_{\text{eff}} \left(\sum_i \frac{\partial}{\partial m_i}\frac{\partial \Gamma}{\partial \eta}(x_i - m_i) \right)^2 \rangle +$$

$$+\langle H_{\text{eff}} \left(\sum_\mu \frac{\partial}{\partial f_\mu}\frac{\partial \Gamma}{\partial \eta}(i\hat{h}_\mu - f_\mu) \right)^2 \rangle - 2\langle H_{\text{eff}}^2 \left[\sum_i (x_i - m_i)\frac{\partial \langle H \rangle}{\partial m_i}\rangle + \right.$$

$$\left. + \sum_\mu (i\hat{h}_\mu - f_\mu)\frac{\partial \langle H_{\text{eff}} \rangle}{\partial f_\mu} \right]\rangle =$$

$$= \langle H_{\text{eff}}^3 \rangle + \langle H_{\text{eff}} \rangle \langle H_{\text{eff}}^2 \rangle - 2\langle H_{\text{eff}} \rangle^3 + \langle H_{\text{eff}} \rangle \langle \sum_{ij,\mu\nu} \frac{\xi_i^\mu \xi_j^\nu}{N} x_i(x_j - m_j)i\hat{h}_\mu f_\nu\rangle +$$

$$+\langle H_{\text{eff}} \rangle \langle \sum_{ij,\mu\nu} \frac{\xi_i^\mu \xi_j^\nu}{N} x_i m_j i\hat{h}_\mu(i\hat{h}_\nu - f_\nu)\rangle + \langle H_{\text{eff}} \left(\sum_i (x_i - m_i) \sum_\mu \left(-\frac{\xi_i^\mu f_\mu}{\sqrt{N}}\right) \right)^2 \rangle +$$

$$+\langle H_{\text{eff}} \left(\sum_\mu (i\hat{h}_\mu - f_\mu) \sum_i \left(-\frac{\xi_i^\mu m_i}{\sqrt{N}}\right) \right)^2 \rangle +$$

$$-2\langle H_{\text{eff}}^2 \left[\sum_i (x_i - m_i) \sum_\mu \frac{\xi_i^\mu f_\mu}{\sqrt{N}}\rangle + \sum_\mu (i\hat{h}_\mu - f_\mu) \sum_i \frac{\xi_i^\mu m_i}{\sqrt{N}} \right] .$$

$$\text{(A.9)}$$

Since off-diagonal terms do not contribute to the free energy and we have to take care of diagonal terms only, the above expression, computed for a generic coupling η that will be eventually set to 1, reduces to:

$$\frac{\partial^3 \Gamma}{\partial \eta^3} = \langle H_{\text{eff}}^3 \rangle + \langle H_{\text{eff}} \rangle \langle H_{\text{eff}}^2 \rangle - 2\langle H_{\text{eff}} \rangle^3 - \langle H_{\text{eff}} \rangle \alpha N r(1-q) +$$

$$-\langle H_{\text{eff}} \rangle \alpha N q(\tilde{r} - r) + \langle H_{\text{eff}} \left(-\sum_{i,\mu} \frac{\delta x_i}{\sqrt{N}} \xi_i^\mu f_\mu \right)^2 \rangle + \langle H_{\text{eff}} \left(-\sum_{i,\mu} \frac{\delta f_\mu}{\sqrt{N}} \xi_i^\mu m_i \right)^2 \rangle +$$

$$-2\langle H_{\text{eff}}^2 \left(\sum_i \delta x_i \sum_\mu \frac{\xi_i^\mu f_\mu}{\sqrt{N}} + \sum_\mu \delta f_\mu \sum_i \frac{\xi_i^\mu m_i}{\sqrt{N}} \right) .$$

$$\text{(A.10)}$$

We used the shorthand notations $(x_i - m_i) = \delta x_i$ and $(i\hat{h}_\mu - f_\mu) = \delta f_\mu$ for the relative deviations from their mean values. From this, one can prove that in the jamming limit, i.e. as $q \to 1$, higher-order corrections do not contribute. Then the effective potential reduces to Eq. (4.75), which is valid for whatever value of the control parameters, provided that the system is analyzed in the hard-sphere regime. We nevertheless remind that the effective potential is specifically modified in the critical region, close to the jamming threshold, where a leading logarithmic interaction, as a function of the average gaps, takes place. One might wonder why these corrections are so interesting in the physics of jamming. Beyond fully-connected topologies, where they do not play any role, they can nevertheless be of great interest in accounting for finite-dimensional systems. Moreover, this computation allows us to remark once more the connections between the jamming transition, at which these corrections vanish, and a mean-field picture. The jamming point should be therefore regarded as a reasonable critical point around which performing a perturbative expansion in the inverse of the dimension.

References

1. Georges A, Yedidia JS (1991) How to expand around mean-field theory using high-temperature expansions. J Phys A: Math Gen 24:2173
2. Itzykson C, Drouffe J-M (1991) Statistical field theory: Volume 2, strong coupling, Monte Carlo methods, conformal field theory and random systems. Vol 2. Cambridge University Press
3. Le Bellac M (1991) Quantum and statistical field theory. Clarendon Press
4. Nakanishi K, Takayama H (1997) Mean-field theory for a spin-glass model of neural networks: TAP free energy and the paramagnetic to spin-glass transition. J Phys A: Math Gen 30(23):8085

Appendix B
Computation of the Replicon Mode in a High-Dimensional Model of Critical Ecosystems

In Chap. 6, we argued that a direct connection between ecological settings and properties of marginal stability in glasses exist. To corroborate this argument we will enter into more detail of our computation, based on a RS hypothesis. Basically, we need to know the general expression of the partition function and to compute the Hessian matrix obtained by varying to the second order the replicated entropy with respect to the overlap matrix Q_{ab} between rescaled availabilities of resources. In Sect. 6.5 we introduced the following partition function according to [3]:

$$Z = \int_{-\infty}^{N} \prod_i \frac{dg_i}{N} e^{\beta \tilde{F}(\{g_i\})} \prod_{\mu=1}^{P} \int d\Delta_\mu \theta(-\Delta_\mu)\delta\left(\Delta_\mu + \epsilon x_\mu + \frac{1}{N}\sum_i g_i \sigma_{\mu i}\right),$$

(B.1)

which, exponentiating the delta-function, becomes:

$$Z = \int_{-\infty}^{N} \prod_i \frac{dg_i}{N} e^{\beta \tilde{F}(\{g_i\})} \prod_{\mu=1}^{P} \int \frac{d\Delta_\mu d\hat{\Delta}_\mu}{2\pi} \theta(-\Delta_\mu) \; e^{\left[i\sum_\mu \hat{\Delta}_\mu (\Delta_\mu + \epsilon x_\mu + \frac{1}{N}\sum_i g_i \sigma_{\mu i})\right]}.$$

(B.2)

The average over the disorder, i.e. over $\sigma_{\mu i}$ and x_μ, can be done independently. More precisely, x_μ is a Gaussian variable and the average over it immediately leads to $\exp\left[\frac{1}{2}\epsilon^2 \sum_\mu \left(\sum_a \hat{\Delta}_\mu^a\right)^2\right]$. We remind, instead, that the metabolic strategies $\sigma_{\mu i}$ take values $\{0, 1\}$ with probabilities $1 - p$ and p respectively. Under these assumptions, we can safely write the replicated partition function, as a function of the order parameters of the model and their Lagrange multipliers. We define the following auxiliary variables that allow us to make use of the replica method and decouple indices μ and i:

$$m^a \equiv \frac{1}{N}\sum_i^N g_i^a, \qquad Q^{ab} \equiv \frac{1}{N^2}\sum_i^N g_i^a g_i^b.$$

(B.3)

from which Eq. (B.1) becomes:

© Springer Nature Switzerland AG 2019
A. Altieri, *Jamming and Glass Transitions*, Springer Theses,
https://doi.org/10.1007/978-3-030-23600-7

$$\overline{Z^n} = \int \prod_{a \le b} \frac{dQ^{ab} d\hat{Q}^{ab}}{2\pi} \int \prod_a \frac{dm^a d\hat{m}^a}{2\pi} \exp\left[i \sum_{a \le b} Q^{ab} \hat{Q}^{ab} + i \sum_a \hat{m}^a m^a \right] \times$$

$$\prod_i \left\{ \int \prod_a \frac{dg_i^a}{N} \exp\left[\sum_a \beta \tilde{F}_i(g_i^a) - \frac{i}{N} \sum_a \hat{m}^a g_i^a - \frac{i}{N^2} \sum_{a \le b} \hat{Q}^{ab} g_i^a g_i^b \right] \right\} \times$$

$$\prod_\mu \left\{ \int \prod_a \frac{d\Delta_\mu^a d\hat{\Delta}_\mu^a}{2\pi} \prod_a \theta(-\Delta_\mu^a) \exp\left[i \sum_a \hat{\Delta}_\mu^a (\Delta_\mu^a + pm^a) + \right.\right.$$

$$\left.\left. - \frac{1}{2} \sum_{a,b} \left(p(1-p)Q^{ab} + \epsilon^2\right) \hat{\Delta}_\mu^a \hat{\Delta}_\mu^b \right] \right\} \right\}$$

(B.4)

where $g_i \equiv N(1 - h_i)$ is an appropriate rescaling of the availabilities h_i. The action S, as a function only of the quantities subject to a saddle-point computation, can be rewritten as [1]:

$$S \propto \left[i \sum_{a \le b} Q^{ab} \hat{Q}^{ab} + i \sum_a \hat{m}^a m^a - \frac{i}{N} \sum_i \sum_a \hat{m}^a g_i^a - \frac{i}{N^2} \sum_i \sum_{a \le b} \hat{Q}^{ab} g_i^a g_i^b + \right.$$

$$\left. + i \sum_\mu \sum_a \hat{\Delta}_\mu^a (\Delta_\mu^a + pm^a) - \frac{1}{2} \sum_\mu \sum_{a,b} \left(p(1-p)Q^{ab} + \epsilon^2\right) \hat{\Delta}_\mu^a \hat{\Delta}_\mu^b \right].$$

(B.5)

All the variables labeled by $\hat{\ }$ stand for Lagrange multipliers, each associated with its corresponding variable without $\hat{\ }$. The relevant terms to focus on are the following ones, where we denote as $\langle \bullet \rangle_c$ the connected correlation function:

$$\frac{\partial S}{\partial Q_{ab}} = i\hat{Q}_{ab} + P\left(-\frac{1}{2}p(1-p)\langle \hat{\Delta}_a \hat{\Delta}_b \rangle_c\right)$$

$$\frac{\partial^2 S}{\partial Q_{ab} \partial Q_{cd}} = P^2 \frac{1}{4} p^2(1-p)^2 \langle \hat{\Delta}_a \hat{\Delta}_b, \hat{\Delta}_c \hat{\Delta}_d \rangle_c$$

$$\frac{\partial S}{\partial \hat{Q}_{ab}} = iQ_{ab} - \frac{i}{N}\langle g_a g_b \rangle_c$$

$$\frac{\partial^2 S}{\partial \hat{Q}_{ab} \partial \hat{Q}_{cd}} = \frac{1}{N^2}\langle g_a g_b, g_c g_d \rangle_c$$

(B.6)

$$\frac{\partial S}{\partial m_a} = i\hat{m}_a + Pip\langle \hat{\Delta}_a \rangle$$

$$\frac{\partial^2 S}{\partial m_a \partial m_b} = -P^2 p^2 \langle \hat{\Delta}_a \hat{\Delta}_b \rangle_c$$

$$\frac{\partial S}{\partial \hat{m}_a} = im_a - i\langle g_a \rangle$$

$$\frac{\partial^2 S}{\partial \hat{m}_a \partial \hat{m}_b} = \langle g_a g_b \rangle_c \ .$$

We did not show terms like $\frac{\partial^2 S}{\partial Q_{ab}\partial \hat{Q}_{cd}} = \frac{\partial^2 S}{\partial \hat{Q}_{ab}\partial Q_{cd}} = \frac{\partial^2 S}{\partial m_a \partial \hat{m}_b} = \frac{\partial^2 S}{\partial \hat{m}_a \partial m_b}$ that provide a constant factor only. They are nevertheless essential to obtain the full expression of the Hessian matrix. We show in the following the expression for the variation of the action S with respect to the overlap matrix Q_{ab}. It consists of four terms, i.e.:

$$\frac{dS}{dQ_{ab}} = \frac{\partial S}{\partial Q_{ab}} + \sum_{ef}\frac{\partial S}{\partial \hat{Q}_{ef}}\frac{d\hat{Q}_{ef}}{dQ_{ab}} + \sum_{e}\frac{\partial S}{\partial m_e}\frac{dm_e}{dQ_{ab}} + \sum_{e}\frac{\partial S}{\partial \hat{m}_e}\frac{d\hat{m}_e}{dQ_{ab}} \,. \quad (B.7)$$

In principle, the computation of the second total derivative looks rather involved, as one can notice from the expression below:

$$\begin{aligned}
\frac{d^2 S}{dQ_{ab}dQ_{cd}} =\ & \frac{\partial^2 S}{\partial Q_{ab}\partial Q_{cd}} + \sum_{ef}\frac{\partial^2 S}{\partial Q_{ab}\partial \hat{Q}_{ef}}\frac{d\hat{Q}_{ef}}{dQ_{cd}} + \sum_{ef}\frac{\partial^2 S}{\partial Q_{cd}\partial \hat{Q}_{ef}}\frac{d\hat{Q}_{ef}}{dQ_{ab}} + \\
& + \sum_{ef}\frac{d^2\hat{Q}_{ef}}{dQ_{ab}dQ_{cd}}\frac{\partial S}{\partial \hat{Q}_{ef}} + \sum_{ef,gh}\frac{\partial^2 S}{\partial \hat{Q}_{ef}\partial \hat{Q}_{gh}}\frac{d\hat{Q}_{ef}}{dQ_{ab}}\frac{d\hat{Q}_{gh}}{dQ_{cd}} + \\
& + \sum_{ef,gh}\frac{\partial^2 S}{\partial \hat{Q}_{ef}\partial \hat{Q}_{gh}}\frac{d\hat{Q}_{ef}}{dQ_{cd}}\frac{d\hat{Q}_{gh}}{dQ_{ab}} + \sum_{e}\frac{\partial^2 S}{\partial Q_{ab}\partial m_e}\frac{dm_e}{dQ_{cd}} + \\
& + \sum_{e}\frac{\partial^2 S}{\partial Q_{cd}\partial m_e}\frac{dm_e}{dQ_{ab}} + \sum_{ef}\frac{\partial^2 S}{\partial m_e\partial m_f}\frac{dm_e}{dQ_{ab}}\frac{dm_f}{dQ_{cd}} + \\
& + \sum_{e}\frac{d^2 m_e}{dQ_{ab}dQ_{cd}}\frac{\partial S}{\partial m_e} + \sum_{e}\frac{\partial^2 S}{\partial Q_{ab}\partial \hat{m}_e}\frac{d\hat{m}_e}{dQ_{cd}} + \\
& + \sum_{e}\frac{\partial^2 S}{\partial Q_{cd}\partial \hat{m}_e}\frac{d\hat{m}_e}{dQ_{ab}} + \sum_{ef}\frac{\partial^2 S}{\partial \hat{m}_e\partial \hat{m}_f}\frac{d\hat{m}_e}{dQ_{ab}}\frac{d\hat{m}_f}{dQ_{cd}} + \\
& + \sum_{e}\frac{d^2 \hat{m}_e}{dQ_{ab}dQ_{cd}}\frac{\partial S}{\partial \hat{m}_e} \,.
\end{aligned}$$

$$(B.8)$$

We can nevertheless take advantage of the fact that several terms, coming from mixed partial derivatives, cancel out with each other. Furthermore, thanks to the saddle-point condition with respect to m, we can use this additional relation:

$$\frac{d}{dQ_{ab}}\frac{\partial S}{\partial m_c} = \frac{\partial^2 S}{\partial m_c\partial Q_{ab}} + \frac{\partial^2 S}{\partial m_c^2}\frac{dm_c}{dQ_{ab}} = 0 \,, \quad (B.9)$$

which allows us to further simplify the computation in terms of dm_c/dQ_{ab}. The expression for the stability matrix then reads:

$$\frac{d^2 S}{dQ_{ab}dQ_{cd}} = \frac{\partial^2 S}{\partial Q_{ab}\partial Q_{cd}} + \sum_{ef} \frac{\partial^2 S}{\partial Q_{ab}\partial \hat{Q}_{ef}} \frac{d\hat{Q}_{ef}}{dQ_{cd}} + \sum_{ef} \frac{\partial^2 S}{\partial Q_{cd}\partial \hat{Q}_{ef}} \frac{d\hat{Q}_{ef}}{dQ_{ab}} +$$

$$+ \sum_{ef} \frac{d^2 \hat{Q}_{ef}}{dQ_{ab}dQ_{cd}} \frac{\partial S}{\partial \hat{Q}_{ef}} + \sum_{ef,gh} \frac{\partial^2 S}{\partial \hat{Q}_{ef}\partial \hat{Q}_{gh}} \frac{d\hat{Q}_{ef}}{dQ_{ab}} \frac{d\hat{Q}_{gh}}{dQ_{cd}} +$$

$$+ \sum_{ef,gh} \frac{\partial^2 S}{\partial \hat{Q}_{ef}\partial \hat{Q}_{gh}} \frac{d\hat{Q}_{ef}}{dQ_{cd}} \frac{d\hat{Q}_{gh}}{dQ_{ab}} ,$$

$$(B.10)$$

which in other terms implies:

$$\frac{d^2 S}{dQ_{ab}dQ_{cd}} = \frac{\partial^2 S}{\partial Q_{ab}\partial Q_{cd}} - \sum_{ef} \frac{\partial^2 S}{\partial Q_{ab}\partial \hat{Q}_{ef}} \frac{\partial^2 S}{\partial \hat{Q}_{ef}\partial Q_{cd}} \left(\frac{\partial^2 S}{\partial \hat{Q}_{ef}^2} \right)^{-1} +$$

$$- \sum_{ef} \frac{\partial^2 S}{\partial Q_{cd}\partial \hat{Q}_{ef}} \frac{\partial^2 S}{\partial \hat{Q}_{ef}\partial Q_{ab}} \left(\frac{\partial^2 S}{\partial \hat{Q}_{ef}^2} \right)^{-1} + \sum_{ef,gh} \frac{\partial^2 S}{\partial \hat{Q}_{ef}\partial \hat{Q}_{gh}} \frac{d\hat{Q}_{ef}}{dQ_{ab}} \frac{d\hat{Q}_{gh}}{dQ_{cd}} +$$

$$+ \sum_{ef,gh} \frac{\partial^2 S}{\partial \hat{Q}_{ef}\partial \hat{Q}_{gh}} \frac{d\hat{Q}_{ef}}{dQ_{cd}} \frac{d\hat{Q}_{gh}}{dQ_{ab}} .$$

$$(B.11)$$

The resulting expression, taking also mixing terms into account, might appear rather involved. However, if $(ef) = (gh)$, the only contribution to take into account is that due to different couples of replica indices. Then, gathering all relevant contributions together, we end up with this expression:

$$\frac{d^2 S}{dQ_{ab}dQ_{cd}} = \frac{P^2}{4} p^2 (1-p)^2 \langle \hat{\Delta}_a \hat{\Delta}_b, \hat{\Delta}_c \hat{\Delta}_d \rangle_c - \frac{2N^2 \langle g_a g_b, g_c g_d \rangle_c}{\langle g_a^2 g_b^2 \rangle_c \langle g_c^2 g_d^2 \rangle_c} , \qquad (B.12)$$

where the two terms have a dimensional dependence proportional to P^2 and N^2 respectively, and each g is $O(\sqrt{N})$. However, we should expect a sub-dominant contribution from the second kind of correlator in the thermodynamic limit. Indeed, once a Gaussian auxiliary variable is introduced to decouple replicas, the only term that depends on z is negligible in the thermodynamic limit and it can be crossed out. If the rescaled variables in z have a subleading contribution, we do not have to take care of the second g-dependent correlator. We should check this assumption more accurately, via numerical tests. For the moment we focus on the diagonal term, that is the partial derivative with respect to Q_{ab} and Q_{cd}:

$$\frac{\partial^2 S}{\partial Q_{ab}\partial Q_{cd}} = \frac{P^2}{4} p^2 (1-p)^2 \langle \hat{\Delta}_a \hat{\Delta}_b, \hat{\Delta}_c \hat{\Delta}_d \rangle_c \equiv \mathcal{M}_{ab,cd} , \qquad (B.13)$$

where we introduced the $\mathcal{M}_{ab,cd}$ matrix, usually defined as *mass matrix* in field theory. According to the replica symmetry, the $\mathcal{M}_{ab,cd}$ matrix can be diagonalized as:

$$\mathcal{M}_{ab,cd} = M_{ab,ab} \left(\frac{\delta_{ac}\delta_{bd} + \delta_{ad}\delta_{bc}}{2} \right) + M_{ab,ac} \left(\frac{\delta_{ac} + \delta_{bd} + \delta_{ad} + \delta_{bc}}{4} \right) + M_{ab,cd} \tag{B.14}$$

and its projection on the replicon subspace yields [2]:

$$\lambda_{\text{repl}} \equiv M_{ab,ab} - 2M_{ab,bc} + M_{ab,cd} =$$
$$= \frac{P^2 p^2 (1-p)^2}{4} \left[\langle \hat{\Delta}_a^2 \hat{\Delta}_b^2 \rangle - 2 \langle \hat{\Delta}_a^2 \hat{\Delta}_b \hat{\Delta}_c \rangle + \langle \hat{\Delta}_a \hat{\Delta}_b \hat{\Delta}_c \hat{\Delta}_d \rangle \right] . \tag{B.15}$$

Let us see a smart way to rewrite the correlator, first focusing on that with all different replica indices, i.e. $\langle \hat{\Delta}_a \hat{\Delta}_b \hat{\Delta}_c \hat{\Delta}_d \rangle$. It reads:

$$= \frac{\int \prod_a \frac{d\Delta_a d\hat{\Delta}_a}{2\pi} \theta(-\Delta_a) e^{\left[i \sum_a \hat{\Delta}_a (\Delta_a + pm^*) - \frac{p(1-p)(\bar{q}-q)}{2} \hat{\Delta}_a^2 - \frac{p(1-p)q+\epsilon^2}{2} (\sum_a \hat{\Delta}_a)^2 \right]} \hat{\Delta}_a \hat{\Delta}_b \hat{\Delta}_c \hat{\Delta}_d}{\int \prod_a \frac{d\Delta_a d\hat{\Delta}_a}{2\pi} \theta(-\Delta_a) e^{\left[i \sum_a \hat{\Delta}_a (\Delta_a + pm^*) - \frac{p(1-p)(\bar{q}-q)}{2} \hat{\Delta}_a^2 - \frac{p(1-p)q+\epsilon^2}{2} (\sum_a \hat{\Delta}_a)^2 \right]}} . \tag{B.16}$$

At this point we can introduce an auxiliary Gaussian variable z to make all replicas uncorrelated. This allows us to split the quadratic terms in the following way:

$$\langle \hat{\Delta}_a \hat{\Delta}_b \hat{\Delta}_c \hat{\Delta}_d \rangle = \int \mathcal{D}z \frac{\left(\int \frac{d\Delta d\hat{\Delta}}{2\pi} \theta(-\Delta) e^{i\hat{\Delta}(\Delta+pm^*) - \frac{p(1-p)(\bar{q}-q)}{2} \hat{\Delta}^2 + iz\sqrt{p(1-p)q+\epsilon^2}\hat{\Delta}} \hat{\Delta} \right)^4}{\left(\int \frac{d\Delta d\hat{\Delta}}{2\pi} \theta(-\Delta) e^{i\hat{\Delta}(\Delta+pm^*) - \frac{p(1-p)(\bar{q}-q)}{2} \hat{\Delta}^2 + iz\sqrt{p(1-p)q+\epsilon^2}\hat{\Delta}} \right)^4} \tag{B.17}$$

where this expression represents the expectation value of the $\hat{\Delta}$-correlator conditioned to z, then averaged over the Gaussian measure. Other correlators, appearing in Eq. (B.15), are:

$$\langle \hat{\Delta}_a^2 \hat{\Delta}_b^2 \rangle = \frac{\int \prod_a \frac{d\Delta_a d\hat{\Delta}_a}{2\pi} \theta(-\Delta_a) e^{\left[i \sum_a \hat{\Delta}_a (\Delta_a + pm^*) - \frac{p(1-p)(\bar{q}-q)}{2} \hat{\Delta}_a^2 - \frac{p(1-p)q+\epsilon^2}{2} (\sum_a \hat{\Delta}_a)^2 \right]} \hat{\Delta}_a^2 \hat{\Delta}_b^2}{\int \prod_a \frac{d\Delta_a d\hat{\Delta}_a}{2\pi} \theta(-\Delta_a) e^{\left[i \sum_a \hat{\Delta}_a (\Delta_a + pm^*) - \frac{p(1-p)(\bar{q}-q)}{2} \hat{\Delta}_a^2 - \frac{p(1-p)q+\epsilon^2}{2} (\sum_a \hat{\Delta}_a)^2 \right]}}$$
$$= \int \mathcal{D}z \frac{\left(\int \frac{d\Delta d\hat{\Delta}}{2\pi} \theta(-\Delta) e^{i\hat{\Delta}(\Delta+pm^*) - \frac{p(1-p)(\bar{q}-q)}{2} \hat{\Delta}^2 + iz\sqrt{p(1-p)q+\epsilon^2}\hat{\Delta}} \hat{\Delta}^2 \right)^2}{\left(\int \frac{d\Delta d\hat{\Delta}}{2\pi} \theta(-\Delta) e^{i\hat{\Delta}(\Delta+pm^*) - \frac{p(1-p)(\bar{q}-q)}{2} \hat{\Delta}^2 + iz\sqrt{p(1-p)q+\epsilon^2}\hat{\Delta}} \right)^4} \tag{B.18}$$

and $\langle \hat{\Delta}_a^2 \hat{\Delta}_b \hat{\Delta}_c \rangle$:

$$= \frac{\int \prod_a \frac{d\Delta_a d\hat{\Delta}_a}{2\pi} \theta(-\Delta_a) e^{\left[i \sum_a \hat{\Delta}_a (\Delta_a + pm^*) - \frac{p(1-p)(\bar{q}-q)}{2} \hat{\Delta}_a^2 - \frac{p(1-p)q+\epsilon^2}{2} (\sum_a \hat{\Delta}_a)^2 \right]} \hat{\Delta}_a^2 \hat{\Delta}_b \hat{\Delta}_c}{\int \prod_a \frac{d\Delta_a d\hat{\Delta}_a}{2\pi} \theta(-\Delta_a) e^{\left[i \sum_a \hat{\Delta}_a (\Delta_a + pm^*) - \frac{p(1-p)(\bar{q}-q)}{2} \hat{\Delta}_a^2 - \frac{p(1-p)q+\epsilon^2}{2} (\sum_a \hat{\Delta}_a)^2 \right]}} . \tag{B.19}$$

The same treatment can be done in the above expression, splitting the whole integral in two contributions, one depending on $\hat{\Delta}^2$ and another on $\hat{\Delta}$. Again, we use the trick of introducing a Gaussian variable to decouple replicas and we safely repeat the same procedure as in Eq. (B.17). If we consider the first power of the numerator N appearing in Eq. (B.17), we can immediately integrate over $\hat{\Delta}$ and obtain:

$$
N = \int \frac{d\Delta \theta(-\Delta)}{\sqrt{2\pi}} \frac{e^{-\frac{(\Delta+pm^*+z\sqrt{p(1-p)q+\epsilon^2})^2}{2p(1-p)(\bar{q}-q)}} \left(i\,(\Delta + pm^*) + iz\sqrt{p(1-p)q+\epsilon^2} \right)}{[p(1-p)(\bar{q}-q)]^{3/2}} =
$$

$$
= -i\, \frac{e^{-\frac{(m^*p+z\sqrt{p(1-p)q+\epsilon^2})^2}{2p(1-p)(\bar{q}-q)}}}{\sqrt{p(1-p)(\bar{q}-q)}} \,.
$$

(B.20)

Computing any single term step by step could be rather involved. We can however notice that the three correlators contributing to the replicon mode are appropriate combinations of the common denominator D, as shown in the Supplementary Material of [1]:

$$
D = \int \frac{d\Delta \theta(-\Delta)}{\sqrt{2\pi}} \frac{e^{-\frac{(\Delta+pm^*+z\sqrt{p(1-p)q+\epsilon^2})^2}{2p(1-p)(\bar{q}-q)}}}{\sqrt{p(1-p)(\bar{q}-q)}} = \frac{1}{2}\left\{ 1 + \mathrm{Erf}\left(\frac{m^*p + z\sqrt{p(1-p)q+\epsilon^2}}{\sqrt{2}\sqrt{p(1-p)(\bar{q}-q)}} \right) \right\} \,.
$$

(B.21)

Then the final combination, to be still integrated over z, reads:

$$
\langle \hat{\Delta}_a^2 \hat{\Delta}_b^2 \rangle_z - 2\langle \hat{\Delta}_a^2 \hat{\Delta}_b \hat{\Delta}_c \rangle_z + \langle \hat{\Delta}_a \hat{\Delta}_b \hat{\Delta}_c \hat{\Delta}_d \rangle_z = \left(\frac{\partial^2 D}{\partial z^2} \frac{1}{\psi^2 i^2 D^2} \right)^2 +
$$

$$
- 2 \left(\frac{\partial^2 D}{\partial z^2} \frac{1}{\psi^2 i^2 D^2} \right) \left(\frac{\partial D}{\partial z} \frac{1}{\psi i D} \right)^2 + \left(\frac{\partial D}{\partial z} \frac{1}{\psi i D} \right)^4 ,
$$

(B.22)

where we introduced the notations:

$$
\psi = \sqrt{p(1-p)q+\epsilon^2}\,, \qquad b = p(1-p)(\bar{q}-q)\,, \qquad c = pm^* \,. \qquad \text{(B.23)}
$$

With reference to Eqs. (B.22) and (B.23), the replicon mode turns out to be:

$$
\lambda_{\mathrm{repl}} \propto \int Dz \frac{4e^{-\frac{2(c+\psi z)^2}{b}} \left[b + 2\sqrt{2\pi b}(c+\psi z)e^{\frac{(c+\psi z)^2}{2b}} + 2\pi(c+\psi z)^2 e^{\frac{(c+\psi z)^2}{b}} \right]}{b^3 \pi^2 \left[1 + \mathrm{Erf}\left(\frac{c+\psi z}{\sqrt{2b}} \right) \right]^4} \,.
$$

(B.24)

Except for certain specific limits, its exact determination might be rather difficult, as the error function also depends on z. Let us make some general comments on it. In the main text we limited the analysis to positive values of the parameters ψ, b, c. However, it is immediate to see that when c becomes negative, the denominator diverges and might generate unsafe behaviors in the integrating function. The case for negative c should be considered carefully, albeit it is not of interest for the shielded phase.

We evaluated numerically the integral (B.24) and analyzed whether different phases can appear for finite ψ and for $\psi \to 0$, respectively. An interesting aspect to point out is the decreasing behavior of the replicon mode when both ψ and b tend to zero, as a signature of the approaching S phase.

References

1. Altieri A, Franz S (2019) Constraint satisfaction mechanisms for marginal stability and criticality in large ecosystems. Phys Rev E 99(1):010401(R)
2. De Dominicis C, Giardina I (2006) Random fields and spin glasses. Cambridge University Press
3. Tikhonov M, Monasson R (2017) Collective phase in resource competition in a highly diverse ecosystem. Phys Rev Lett 118(4):048103

Appendix C
Diagrammatic Rules for the M-Layer Construction in the Bethe Approximation

We consider a generic graph $\mathcal{G}$, according to which we can define the following rules:
On each edge $e \equiv i \rightarrow j$

- $\hat{B}_{\lambda_e}(k_e; \ \mu_e^i, \mu_e^j)$
- $g_{\lambda_e}(\sigma_i) \, g_{\lambda_e}(\sigma_j)$
- $\sum_{\lambda_e}$
- $\int_{[-\pi,\pi]^d} \frac{k_e}{(2\pi)^d}$

On each vertex i

- $P_{z-z_i}^{Bethe}(\sigma_i)$
- $(2\pi)^d \, \tilde{\delta} \left(\sum_{j \in \partial_i^{in}} k_{j \rightarrow i} - \sum_{j \in \partial_i^{out}} k_{i \rightarrow j} \right)$
- $\mathbb{I}\left(\{|\mu_e^i|\}_e \text{ pairwise different} \right)$
- $\sum_{\{\mu_e^i\}_e}$
- $\int \sigma_i$

The Fourier transform of the generating function of non-backtracking random walks has the form:

$$\hat{B}_\lambda(k; \mu, \alpha) = c_\lambda^{\mu\alpha}(k) \, \frac{1}{1 + (2d-1)\lambda^2 - 2\lambda \sum_{\mu=1}^d \cos(k_\mu)} \cdot \tag{C.1}$$

Considering $\lambda_c = 1/(2d-1)$, small $\tau = \lambda_c - \lambda$, small k^2, the singular behaviour is given by:

$$\hat{B}_\lambda(k; \mu, \alpha) \sim c_{\lambda_c}^{\mu\alpha}(0) \, \frac{1}{(2d-2)\tau + \frac{1}{2d-1}k^2 - \tau k^2} \,, \tag{C.2}$$

where the prefactor is $c_{\lambda_c}^{\mu\alpha}(0) = (d-1)/d(2d-1)^2$.

© Springer Nature Switzerland AG 2019
A. Altieri, *Jamming and Glass Transitions*, Springer Theses,
https://doi.org/10.1007/978-3-030-23600-7

C.0.1   Non-backtracking Walks

We aim to extend some results by Fitzner and van der Hofstad [1] to enumerate non-backtracking walks (NBWs) on the lattice $\mathbb{Z}^d$, for $d \geq 2$. We shall not consider the case in one dimension, which is trivial.

We first introduce some conventions that we shall adopt throughout this Section. In particular, we label with μ, ν and α the $2d$ directions in the d-dimensional lattice, each index taking values in $\{\pm 1, \pm 2, \ldots, \pm d\}$. The corresponding versors $e_\mu \in \mathbb{Z}^d$ are defined by $(e_\mu)_m \equiv \text{sgn}(\mu)\, \delta_{|\mu|, m}$ for $m = 1, \ldots, d$. For a generic d-dimensional vector k we also define $k_{-|\mu|} \equiv -k_{|\mu|}$.

A NBW of length $n \geq 1$ in $\mathbb{Z}^d$ is a sequence $\omega = (x_0, x_1, \ldots, x_n)$, $x_i \in \mathbb{Z}^d$, such that $\|x_{i+1} - x_i\| = 1$ and $x_{i+2} \neq x_i$. We call α the *start direction* of w, if $x_1 - x_0 = e_\alpha$, and μ the *end direction* of w, if $x_n - x_{n-1} = e_\mu$.

Without any loss of generality we can assume that $x_0 = 0$. Let us call $a_n^\alpha(x)$ the number of NBW of length n in $\mathbb{Z}^d$ starting from the origin in the direction α and ending in x (i.e. $x_n = x$). It is also convenient to define $b_n^{\mu,\alpha}(x)$ as the number of NBW of length n in $\mathbb{Z}^d$ starting from the origin in the direction α and ending in x whose final direction is not μ. These definitions lead to the following relations:

$$b_n^{\mu,\alpha}(x) = \sum_{\nu \neq \mu} b_{n-1}^{-\nu,\alpha}(x - e_\nu) \qquad\qquad n \geq 2 \qquad\qquad \text{(C.3)}$$

$$b_1^{\mu,\alpha}(x) = (1 - \delta_{\mu\alpha})\delta(x - e_\alpha) \qquad\qquad\qquad\qquad \text{(C.4)}$$

and

$$a_n^\alpha(x) = \sum_\nu b_{n-1}^{-\nu,\alpha}(x - e_\nu) \qquad\qquad n \geq 2 \qquad\qquad \text{(C.5)}$$

$$a_n^\alpha(x) = b_n^{\nu,\alpha}(x) + b_{n-1}^{-\nu,\alpha}(x - e_\nu) \qquad\qquad n \geq 2, \forall \nu \qquad\qquad \text{(C.6)}$$

$$a_1^\alpha(x) = \delta(x - e_\alpha). \qquad\qquad\qquad\qquad\qquad \text{(C.7)}$$

For a generic and well-behaved function $f(x), x \in \mathbb{Z}^d$, we shall denote its Fourier transform with $\hat{f}(k)$, $k \in [-\pi, \pi]^d$, so that:

$$\hat{f}(k) = \sum_{x \in \mathbb{Z}^d} f(x)\, e^{ikx} \qquad\qquad\qquad\qquad \text{(C.8)}$$

$$f(x) = \int_{[-\pi,\pi]^d} \frac{k}{(2\pi)^d}\, \hat{f}(k)\, e^{ikx} . \qquad\qquad \text{(C.9)}$$

In Fourier space Eqs. (C.3)–(C.7) imply:

$$\hat{b}_n^{\mu,\alpha}(k) = \sum_{\nu \neq \mu} \hat{b}_{n-1}^{-\nu,\alpha}(k)\, e^{ik_\nu} \qquad\qquad n \geq 2 \qquad\qquad \text{(C.10)}$$

$$\hat{b}_1^{\mu,\alpha}(k) = (1 - \delta_{\mu\alpha})\, e^{ik_\alpha} \qquad\qquad\qquad\qquad \text{(C.11)}$$

and

$$\hat{a}_n^\alpha(k) = \sum_\nu \hat{b}_{n-1}^{-\nu,\alpha}(k)\, e^{ik_\nu} \qquad\qquad n \ge 2 \qquad\qquad \text{(C.12)}$$

$$\hat{a}_n^\alpha(k) = \hat{b}_n^{\nu,\alpha}(k) + \hat{b}_{n-1}^{-\nu,\alpha}(k)\, e^{ik_\nu} \qquad n \ge 2,\, \forall \nu \qquad \text{(C.13)}$$

$$\hat{a}_1^\alpha(k) = e^{ik_\alpha} . \qquad\qquad\qquad\qquad\qquad\qquad \text{(C.14)}$$

We now denote with $\vec{a}_n(k)$ the $2d$-dimensional vector with elements $\hat{a}_n^\alpha(k)$, and with $\hat{\mathbf{b}}_n(k)$ the $2d \times 2d$ matrix with elements $\hat{b}_n^{\mu,\alpha}(k)$. We also indicate with $\vec{1}$ the all ones vector and with $\mathbf{1}$ the all ones matrix, with $\mathbf{I}$ the matrix identity, with $\vec{v}^T$ vector transposition, and introduce the matrices $\mathbf{D}(k)$ and $\mathbf{J}$ with elements:

$$D^{\mu\alpha}(k) = \delta_{\mu,\alpha}\, e^{ik_\mu}, \qquad\qquad\qquad \text{(C.15)}$$

$$J^{\mu\alpha} = \delta_{\mu,-\alpha} . \qquad\qquad\qquad\quad \text{(C.16)}$$

We can then compactly write Eqs. (C.12)–(C.13) as

$$\vec{a}_n^T(k) = \vec{1}^T \mathbf{D}(-k)\hat{\mathbf{b}}_{n-1}(k) \qquad\qquad \text{(C.17)}$$

$$\vec{1}\vec{a}_n^T(k) = \hat{\mathbf{b}}_n(k) + \mathbf{D}(k)\mathbf{J}\hat{\mathbf{b}}_{n-1}(k) . \qquad \text{(C.18)}$$

These recurrent relations can be conveniently solved in terms of generating functions. For a succession $f_n,\, n \ge 1$, we formally define its generating function as:

$$F_z = \sum_{n=1}^{+\infty} f_n\, z^n . \qquad\qquad\qquad \text{(C.19)}$$

Therefore, associating $\vec{A}_z(k)$ to $\vec{a}_n(k)$ and $\mathbf{B}_z(k)$ to $\hat{\mathbf{b}}_n(k)$, we get:

$$\vec{A}_z^T(k) = z\vec{a}_1^T(k) + z\vec{1}^T \mathbf{D}(-k)\hat{\mathbf{B}}_z(k) \qquad\qquad \text{(C.20)}$$

$$\vec{1}\vec{A}_z^T(k) = \hat{\mathbf{B}}_z(k) + z\vec{1}\vec{a}_1^T(k) - z\hat{\mathbf{b}}_1(k) + z\mathbf{D}(k)\mathbf{J}\hat{\mathbf{B}}_z(k). \qquad \text{(C.21)}$$

The last equation leads to:

$$\hat{\mathbf{B}}_z(k) = (\mathbf{I} + z\mathbf{D}(k)\mathbf{J}))^{-1} \left(\vec{1}\vec{A}_z^T(k) - z\vec{1}\vec{a}_1^T(k) + z\mathbf{b}_1(k) \right), \qquad \text{(C.22)}$$

which, inserted in (C.21), gives:

$$\vec{A}_z^T(k) = z\frac{\vec{a}_1^T(k) + z\vec{1}^T \mathbf{D}(-k)\,(\mathbf{I} + z\mathbf{D}(k)\mathbf{J}))^{-1}\left(\mathbf{b}_1(k) - \vec{1}\vec{a}_1^T(k) \right)}{1 - z\,\vec{1}^T \mathbf{D}(-k)\,(\mathbf{I} + z\mathbf{D}(k)\mathbf{J}))^{-1}\,\vec{1}} . \qquad \text{(C.23)}$$

This expression looks particularly convenient and allows us to write:

$$(\mathbf{I} + z\mathbf{D}(k)\mathbf{J}))^{-1} = \frac{1}{1 - z^2}(\mathbf{I} - z\mathbf{D}(k)\mathbf{J}) , \qquad (C.24)$$

easily derivable from the identity $\mathbf{JD}(k)\mathbf{JD}(k) = \mathbf{I}$.

Equation (C.23) contains a scalar denominator and no matrix inversions, so it can be easily computed, i.e.:

$$A_z^\alpha(k) = \frac{z(e^{ik_\alpha} - z)}{1 + z^2(2d - 1) - 2dz D(k)} , \qquad (C.25)$$

with

$$D(k) \equiv \frac{1}{d} \sum_{\mu=1}^{d} \cos(k_\mu) . \qquad (C.26)$$

Using this result in Eq. (C.22), we also obtain:

$$B_z^{\mu,\alpha}(k) = \frac{1}{1 - z^2}\left((1 - ze^{ik_\mu})A_z^\alpha(k) - z(\delta_{\mu,\alpha} e^{ik_\mu} - z\delta_{\mu,-\alpha})\right) . \qquad (C.27)$$

From this equation we can derive our final result for the generating function of the number of NBWs with start and end directions α and μ respectively. We call this generating function $\hat{B}_z(k; \mu, \alpha)$, through a linear transform given by the matrix:

$$S_{\mu\nu} \equiv (\mathbf{1} - \mathbf{I})_{\mu,\nu}^{-1} = \frac{1}{2d - 1} - \delta_{\mu\nu} . \qquad (C.28)$$

Therefore, the final expression is:

$$\hat{B}_z(k; \mu, \alpha) = (S\,\mathbf{B}(k))_{\mu\alpha} = \frac{c_z^{\mu,\alpha}(k)}{1 + z^2(2d - 1) - 2dz D(k)} , \qquad (C.29)$$

where we also defined:

$$c_z^{\mu,\alpha}(k) \equiv \frac{z}{1 - z^2}\Big[z\big((e^{ik_\alpha} - z)(e^{ik_\mu} - z) + \delta_{\mu,-\alpha}(z(2d D(k) + z - 2dz) - 1)\big) + $$
$$- \delta_{\mu\alpha}e^{ik_\alpha}(2d(D(k) - z)z + z^2 - 1)\Big] . $$
$$(C.30)$$

One can check that $\sum_\mu \hat{B}_z(k; \mu, \alpha) = A_z^\alpha(k)$ as in Eq. (C.25) and correctly expected. The generating function for the total number of NBWs, i.e. $\hat{B}_z(k) \equiv \sum_{\mu,\alpha} \hat{B}_z(k; \mu, \alpha)$, is given by the following relation:

$$\hat{B}_z(k) = \frac{2dz(D(k) - z)}{1 + z^2(2d - 1) - 2dzD(k)} ,$$ (C.31)

which differs from that computed in [1] because of the different convention for the number of NBW of length zero, which is assumed to be zero in our case.

Notice that, since $D(k) \sim 1 - k^2/2d$ for small k, all generating functions we considered in this Appendix have a pole in $z = 1/(2d - 1)$ for $k = 0$.

Reference

1. Fitzner R, van der Hofstad R (2013) Non-backtracking random walk. J Stat Phys 150(2):264

Appendix D
The Symmetry Factor

In Chap. 7 we introduced a perturbative expansion in terms of *fat diagrams*. Once the diagrammatic expansion is defined, natural question to ask is about the minimal amount of information needed to identify a given diagram unambiguously. Under a given symmetry, positions and directions of each vertex have to satisfy specific constraints. For instance, if a graph is symmetric under the interchange of two vertices, its positions require an un-ordered couple.

We suppose to label all the vertices and the legs of each vertex. We can then consider a different labeling of the vertices and the legs and ask whether the adjacency matrix of the graph is different from the original one. The number of different labelings giving the same adjacency matrix corresponds to the symmetry group of the graph. Then, to correctly count everything, we need to divide the final expression by the symmetry group factor.

In the case of Feynman diagrams, the Wick theorem provides a rule: if we multiply by the permutation of the vertices and the permutation of the internal lines, the overall factorials, coming from the Taylor expansion, and the factor $1/k!$ in the coupling constant cancel out. Eventually, the normalizing symmetry factor allows to avoid multiple counting.

According to [1, 2], in a real scalar theory the symmetry factor of a generic diagram can be immediately computed as:

$$S = p2^\gamma \prod_{n=2,3,\ldots} (n!)^{\alpha_n} \tag{D.1}$$

where α_n is the number of pairs of vertices connected by n identical self-conjugated lines, γ is the number of lines connecting a vertex to itself and finally p is the number of permutations of vertices keeping the structure of the diagram unchanged for fixed external lines. According to Eq. (D.1), in our case the factors we are looking for are $1/2$ for the *cactus* diagram, i.e. the first order contribution of Fig. 7.4, and $1/6$ for the *Saturn* diagram shown as third example in Fig. 7.4. In the ordinary φ^4 theory, indeed, the symmetry factor of the cactus diagram is given by $S = \frac{4!}{4\cdot3} = 2$, by counting 4 possibilities to connect the internal vertex with the leg on the left side and other 3

© Springer Nature Switzerland AG 2019
A. Altieri, *Jamming and Glass Transitions*, Springer Theses,
https://doi.org/10.1007/978-3-030-23600-7

possibilities to connect it with the leg on the right side, one link being already fixed. In a similar way, for the Saturn diagram, $S = \frac{2(4!)^2}{4 \cdot 4 \cdot 2 \cdot 3!} = 6$, because it has two vertices each contributing with a factor 4!, and there are 4 different ways to attach each external leg to one of the internal vertices and more 3! ways to permute the internal lines with themselves.

We implicitly assume that all the lines are distinguishable. If the diagram is nevertheless sensitive to the direction, its multiplicity depends only on the number of ways needed to construct the same topology. If we generalize this reasoning to our particular case, the symmetry factors are the same as those derived for a generic field theory, provided that only non-backtracking paths are taken into account.

References

1. Cheng T-P, Li L-F (1984) Gauge theory of elementary particle physics. Clarendon Press Oxford
2. Kaku M (1993) Quantum field theory, a modern introduction. Oxford University Press

Printed in the United States
By Bookmasters